System-level Techniques for Analog Performance Enhancement

Bang-Sup Song

System-level Techniques for Analog Performance Enhancement

Springer

Bang-Sup Song
Department of Electrical and Computer Engineering
University of California
San Diego, CA, USA

ISBN 978-3-319-27919-0 ISBN 978-3-319-27921-3 (eBook)
DOI 10.1007/978-3-319-27921-3

Library of Congress Control Number: 2015959480

Printed on acid-free paper

This Springer imprint is published by Springer Nature
The registered company is Springer International Publishing AG Switzerland

Preface

Electronics has undergone drastic changes and shifted towards massive digital signal processing in recent decades. However, even large digital-heavy systems are built around small analog cores, and analog signal processing is still the mainstay of the circuit design. Analog circuits have evolved from old discrete sampled-data circuits to new sophisticated digital-like time-domain switching circuits. In this evolution, numerical design methods relying on numerical analysis have been favored, and simulation tools were given undue credits for that.

Analog circuits and their fundamentals have been improved over the last century even without computer tools. Old analog designs were to improve circuits in relative terms using matching properties. On the contrary, new analog designs are to obtain absolute parameters such as low figure-of-merit, low power, and high speed. After 30 years of refinement, analog designers now come to believe that virtually all analog functions from DC to RF can be integrated at low voltages even with low power. Such euphoria of superior analog performance has been mostly fueled by aggressive device scaling down to the nanometer scale. It is true that simulation-based designs can handle circuits of large complexity, but in an effort to implement analog functions only with high-speed digital switching, the fundamentals of electronics have been often ignored.

More often than not, simulation-based designs end up with somewhat ambiguous and erroneous results ranging from violating the fundamental energy conservation law to obtaining instantaneous small-signal gain during brief large-signal transient period. They have veered off course with no obvious ends, which require necessary corrections. This book raises a concern about such analog design practices and commonly overlooked fallacies. While most recent literatures focus on answering mostly what and how, this book elaborates more on why and how.

Since analog design methodologies are reviewed with specific emphasis on concepts, college-level engineering knowledge would suffice to understand this book. It is written in plain descriptive and illustrative terms with no extensive derivations of equations and simulations so that readers may grasp the essence of this book intuitively without resorting to numerical means. There are seven chapters with examples from DC to RF. After fundamental issues are identified, analog performance-enhancing methods are presented repeatedly using the same principle applicable to the system-level DC servo feedback as well as the simple local feedback.

This book is mainly written and organized to give proper perspectives on the analog designs at all levels. The analog design field is built on collective achievements by numerous contributors, and apologies are extended to those whom the author failed to refer to or recognize correctly in this book.

La Jolla, CA
November 2015

Bang-Sup Song

Contents

Chapter 1
Discrete-Time Switching Circuits

All electronic circuits and systems perform two basic functions: signal generation and detection. Hence their performance or resolution are evaluated by the accuracy signal is handled at a certain speed. Therefore, the objective of all analog designs is to meet specified accuracy and speed requirements with constraints such as supply voltage, power, noise, or signal swing over process, voltage, and temperature (PVT) variations. As CMOS is scaled down to the nanometer range, a new analog design style that relies on switching techniques has emerged and gained momentum.

1.1 Comparators

Most advances in analog designs have been made while designers are constantly searching for better ways to detect weak signals more reliably. The signal detection is a decision-making process. Analog circuits resolve signal with a continuous voltage or current scale while digital circuits represent it with just two levels of high or low values. Since infinite resolution is not possible, the resolution of analog circuits is measured by distortion or nonlinearity, which is defined as an amount of deviation from ideal or linear output value. To represent continuous analog voltages with discrete levels, only a finite number of analog levels should be detected. Comparators perform this level-detection function. That is, the basic function of a comparator is to make a decision on whether an analog input is higher or lower than a reference level. Two main analog issues arise. One is how accurately analog voltage is defined, and the other is how quickly decision is made. Therefore, comparator designs vary widely depending on how these resolution and speed requirements are met.

There exist only three types of analog circuits: Analog amplifier, digital inverter, and analog/digital latch as shown in Fig. 1.1. They are all decision or detection circuits comparing analog inputs to the thresholds of their transfer functions. The threshold is defined as the input voltage that makes the output cross the middle

B.-S. Song, *System-level Techniques for Analog Performance Enhancement*,
DOI 10.1007/978-3-319-27921-3_1

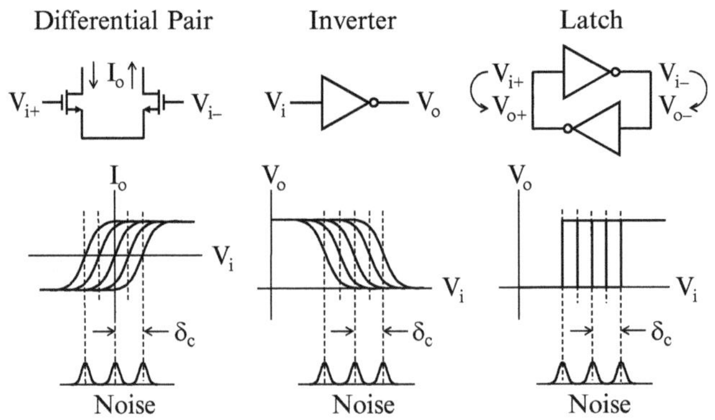

Fig. 1.1 Thresholds in the transfer functions of amplifier, inverter, and latch

point of the output range. Non-Gaussian static DC error also originates from the uncertainty of the threshold like DC offset. The bell-shaped variation of the white Gaussian noise is conceptually added to the thresholds of the transfer functions.

The threshold voltage of the MOS device is the most poorly defined parameter in circuits, and its uncertainty also contributes greatly to the threshold accuracy of the transfer function both in the differential and single-ended forms. In digital circuits, the situation gets aggravated as digital signal swings from rail to rail with hysteresis. The constant and average portion of the threshold shifts is the DC offset, which results from the random mismatch of device sizes and bias conditions. The offset error is considered constant, and contributes only to the systematic offset, which usually inflicts no harm to the system performance. However, the threshold error is non-Gaussian and deterministic with sign and magnitude, and it should be clearly distinguished from the variance of the white Gaussian random noise which lacks sign and magnitude information. Such non-Gaussian errors appear in many different forms in analog circuits and systems, and therefore, reducing them has been the primary design goal of analog designs that demand high performance with fine resolution.

1.1.1 Right, Wrong, or No Decision

If the input of the inverter is high or low for example, the threshold shifts to the right or to the left correspondingly. The same is true to the latch, which is a positive feedback circuit and has a steep transition at the threshold. This hysteresis in the threshold gives rise to the non-Gaussian threshold error marked as δ_c. Unless taken care of, it severely degrades the comparator performance. In all non-resetting sampled-data circuits such as successive-approximation (SAR) ADC, comparator threshold varies widely depending on input data pattern. For the same reason, multi-bit memory is not practical since it is difficult to detect multiple levels reliably.

Although the partial response scheme is used to read and write in hard disks, its multi-level detection scheme is getting sophisticated, and a substantial amount of digital signal processing is involved.

Latch detects a small seed voltage and makes a quick decision to get high and low digital outputs depending on its polarity. Since latch operates with large voltage in switching mode, its hysteresis gets larger, and contributes more to the total non-Gaussian comparator error. It is similar in effect to the offset and threshold shift as it causes comparators to make wrong decisions. A small seed signal sampled on high-impedance latch input is destroyed by coupling or kickback from the latch output. Therefore, latch needs larger seed signal than necessary for its output to flip reliably. Another factor that also contributes to the non-Gaussian sporadic comparator error is metastability, which can occur typically at high clock rates when the latch cannot complete its decision within a given time slot. It often makes large burst-mode errors that cannot be corrected in the digital output during the digital encoding process.

There are three kinds of decision results—right, wrong, and no decisions. As both wrong and no decisions are non-Gaussian and deterministic, they both increase the bit error rate (BER). The wrong decision error results from the decision threshold shift and hysteresis while the no-decision error occurs due to the metastability of the comparator. Note that the wrong decision error can be digitally corrected or reduced by feedback, but the comparator indecision ends up with large digital errors like sparkle codes in the flash ADC, which should be avoided by all means. The non-Gaussian comparator error has been the most critical design constraint in ADC designs. All ADC circuit techniques and systems have been evolved with one single goal in mind to make accurate decisions regardless of this non-Gaussian comparator error. The comparator burst error has often been ignored in recent ADC designs such as in SAR as it occurs sporadically and doesn't degrade the SNR to a great extent.

The same thing can be said about the decisions in digital communications receivers. Wrong decisions can be partly corrected by minimizing mean square errors using digital signal processing schemes such as decision-directed equalization, maximum likelihood decision, error correction, or Viterbi. However, large digital indecision errors cannot be corrected and can degrade the BER.

1.1.2 Non-Gaussian Comparator Error

If this non-Gaussian decision error were not in existence, analog design might have been the easiest engineering practice of all, and memories and storage devices might be using multiple bits per cell by now. However, in reality, they do exist though they cannot be clearly defined. As a result, custom decision circuits should have been developed to meet the requirements of specific analog signal processing and digital communications systems. The history of modern electronics itself reflects the evolutional process that has led to numerous clever ideas, concepts,

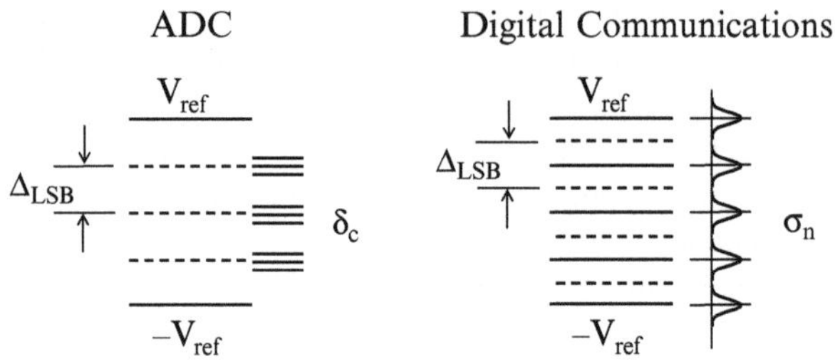

Fig. 1.2 Similarity between the ADC comparator and digital quantizer

and system techniques to remove the comparator errors in decision-making at various stages from the simple latch to the system-level digital maximum-likelihood decision. The bottom line in analog designs is that systems should be configured to always make correct decisions accurately and fast regardless of the non-Gaussian errors of the comparator and latch.

The similarity between the ADC comparator and the bit slicer or quantizer in digital communications receivers is conceptually illustrated in Fig. 1.2. The same minimum resolvable step is given as $\Delta_{\rm LSB}$, and the comparator thresholds are marked with dashed lines. In the ADC, the sampled analog input is compared to the fixed reference levels, but decisions are made with the comparator threshold error $\delta_{\rm c}$. On the other hand, the digital quantizer in digital communications makes decisions on the noisy digital input with a variance of $\sigma_{\rm n}$ since the digital quantizer has no numerical threshold error. Note that assuming the quantization step $\Delta_{\rm LSB}$ is the same, the only difference between them is the fact that the comparator threshold error $\delta_{\rm c}$ is non-Gaussian and deterministic while the white noise variance $\sigma_{\rm n}$ is Gaussian and random. The former results in the comparator error, and similarly increases the BER in decisions as in digital receivers. Like the noise variance affects the BER of the digital quantizer, the comparator error increases a probability of getting erroneous non-Gaussian quantization errors. It is true that the Gaussian portion of the quantization error can be handled like a random noise with a variance. Therefore, there hasn't been any clear distinction between the two terms, quantization error and quantization noise. They are interchangeable in the analysis though the quantization error can be represented with random voltage with sign and magnitude while the random white noise is not.

Unless comparator is reset before every decision, the comparator threshold error $\delta_{\rm c}$ affects the SNR as follows.

$$10\log\frac{S}{Q+\delta_{\rm c}^2} = SQ{\rm NR} - 10\log\left(1+\frac{\delta_{\rm c}^2}{Q}\right). \tag{1.1}$$

For the SNR degradation to be smaller than 1 dB, the threshold error should be smaller than about half the quantization error or a fraction of the quantization step $\Delta_{\rm LSB}$.

$$\delta_c < 0.51 \times \sqrt{Q} = 0.51 \times \sqrt{\frac{\Delta_{\mathrm{LSB}}^2}{12}} \approx 0.15 \times \Delta_{\mathrm{LSB}}. \quad (1.2)$$

This implies that the comparator threshold error should be controlled to be smaller than approximately $\Delta_{\mathrm{LSB}}/6.8$ so that the SNR may not degrade by more than 1 dB.

The noise variance of the signal in digital communications has the same effect on the digital quantizer performance. The low SNR degrades the BER. For the standard BPSK or QPSK with BER = 0.001 (0.1 %), an SNR of about 7 dB is required. From the standard definition of the SNR, we get the following relation.

$$10\log\frac{S}{N} = 10\log\frac{\left(\frac{\Delta_{\mathrm{LSB}}}{2}\right)^2}{2\sigma_{\mathrm{n}}^2} > 7 \ \mathrm{dB}. \quad (1.3)$$

That is, the noise variance should be smaller than a fraction of the quantization step of Δ_{LSB} for the BER to be lower than 0.1 %.

$$\sigma_{\mathrm{n}} < 0.16 \times \Delta_{\mathrm{LSB}}. \quad (1.4)$$

The noise variance should be smaller than approximately $\Delta_{\mathrm{LSB}}/6.3$ so that the BER may not degrade by any more than 0.1 %. The two conditions of (1.2) and (1.4) require about the same level of comparator threshold accuracy required both for ADCs and digital communications receivers. This condition imposes a very stringent requirement on the comparator design that the non-Gaussian comparator error should be suppressed to be way smaller than the minimum quantization step.

As shown in Fig. 1.3, both preamp and latch experience large input swings depending on the previous digital values, and their offsets exhibit hysteresis. Mathematically, the comparator error can be handled like a random variable with a variance of δ_c. Similarly, the quantization error resulting from the ideal uniform quantization steps is non-Gaussian and deterministic, but it is handled like a random variable with a variance. However, the difference between them is that the comparator threshold error is dependent on the seed signal at the preamp input and the digital output of the latch while the quantization error is random and independent of

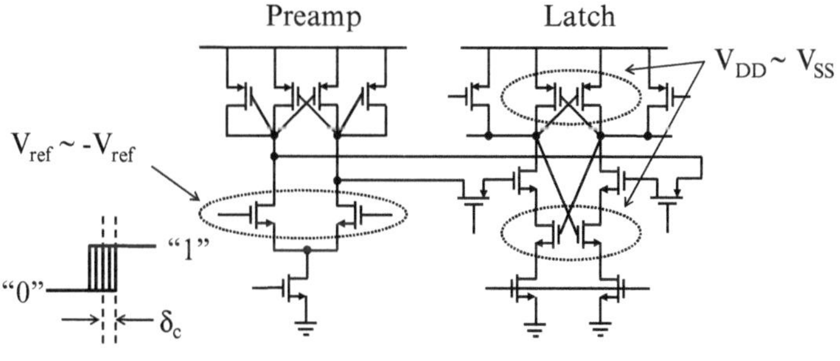

Fig. 1.3 Large voltage stresses on the preamplifier and latch

the input. Therefore, note that the non-Gaussian portion of the comparator error gives rise to the burst-mode but deterministic error, which is closer by nature to the static differential and integral nonlinearities (DNL and INL) in the ADC transfer characteristic than to the Gaussian random noise.

1.1.3 Digital Correction and Feedback

ADCs and DACs used in digital communications systems are required to quantize and to generate analog waveforms with high spurious-free dynamic range (SFDR) and low BER for digital processing. In digital transceivers, either impulse or binary data are commonly shaped using DAC before transmitting, and data received through wireline or wireless channels are quantized using ADC after the channel is adaptively equalized. Since the received data waveform is corrupted by noise in the channels, analog comparator errors in modern digital receivers are designed to be much smaller than the noise variance. Then digital receivers' quantization errors are only limited by the SNR of the received data. That is, digital comparator implements a bit slicer or a quantizer with no comparator error. The digital comparators in digital receivers for BPSK or QPSK (1b), 16-QAM (2b), 64-QAM (3b), 256-QAM (4b), and 1024-QAM (5b) are equivalent to the comparators in ADCs for 1b, 2b, 3b, 4b, and 5b, respectively. To improve the BER further, various sophisticated digital schemes have been developed to date in digital receivers.

The wrong and no decision issues can be addressed both in the analog and the digital domains. In analog comparators, amplifying the seed signal before latching significantly lowers a probability of getting wrong or no decision errors. Preamplifiers experience large input signals, and therefore, need to be reset periodically to alleviate the hysteresis effect. Latches are always stressed by large rail-to-rail signals, and exhibit hysteresis and metastability. Although both preamplifier and latch are reset before decision, sufficient comparator accuracy is not obtained in most cases due to the switch feedthrough, charge injection, and latch metastability. The latch metastability can be avoided by latching twice (double latching), but analog circuit techniques alone cannot reduce the comparator error. To date, two prominent digital techniques have been developed to eliminate the comparator error. One is the digital correction scheme adopted in the pipelined ADC, and the other is the quantizer feedback concept in the oversampling $\Delta\Sigma$ modulator.

In pipelined ADCs, about 3.4 times the comparator error δ_c can be digitally corrected if the redundant range of extra half LSB is covered by overlapping one redundant bit between the pipelined stages. That is, wrong decisions due to the comparator threshold error up to half LSB can be corrected.

$$\delta_c < 3.4 \times (\Delta_{LSB}/6.8) = 0.5 \times \Delta_{LSB}. \tag{1.5}$$

Similarly in $\Delta\Sigma$ modulators, the comparator threshold error can be considered as a part of the quantization error unless it is too large, and its in-band error spectrum is suppressed by the feedback loop gain together with the quantization error.

Note that both pipelined ADCs and $\Delta\Sigma$ modulators allow very large comparator errors since they only resolve a few bits per stage at the most. On the other hand, open-loop ADCs like SAR and flash require a full ADC resolution at every decision still with the comparator error smaller than about 1/6.8 of the LSB ($\Delta_{\rm LSB}/6.8$) if the SNR is not allowed to degrade by any more than 1 dB. This SNR degradation is rarely noticed in the ADC code density test since it is concentrated either at the low end of the spectrum where $1/f$ noise already dominates or spread over the wide range of frequencies due to the nature of sporadic impulse-like broadband error power. The effect can be more readily observed from the time-domain testing of the ADC used for the sparkle code and metastability measurements.

1.1.4 Latch Gain

Analog comparator used in ADCs suffers analog imperfections such as threshold shift, hysteresis, and metastability while digital comparator suffers only the quantization error. Therefore, the analog comparator should be made of a preamplifier followed by a regenerative latch. The regenerative latch is the most numerous circuit used in today's electronics. It is a positive feedback circuit with only two stable states, high or low. No matter what the initial state is, it always settles back to one of two bistable states. Thus the regenerative latch is used as a coarse comparator that finds numerous applications from ADCs to digital memories and registers.

Comparator is a transient circuit, and the initial conditions of the high-impedance input and output are not well defined. Therefore, the time constant at the preamplifier output should be short enough for any transient initial voltage to be amplified fast during the half clock period. If it is too long, the output may still drift away or recover slowly before latching. That is, the input is more vulnerable to the coupling and kickback from the latching operation, which can lead to wrong decision.

The pole locations of the preamplifier and the latch are compared in Fig. 1.4. Both preamplifier and latch need wide bandwidths for proper operation. The transfer function of the preamplifier is

$$H(s) = \frac{g_{\rm m}R}{1 + sRC}, \tag{1.6}$$

which gives a negative real axis pole at $-1/RC$. On the other hand, that of the positive feedback latch is given by

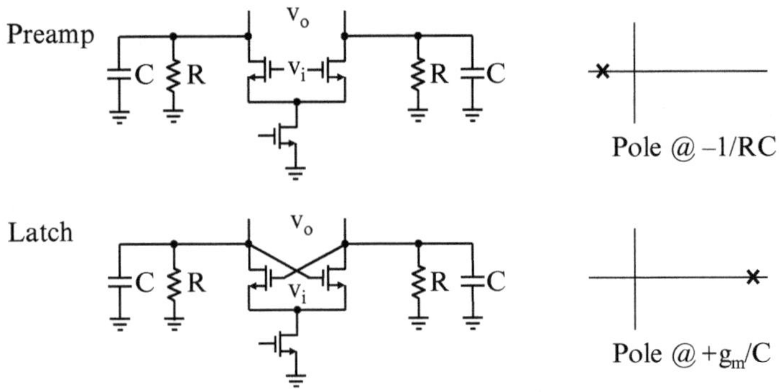

Fig. 1.4 Pole locations in *s*-plane of the preamplifier and latch

Fig. 1.5 Transient responses of the preamplifier and latch

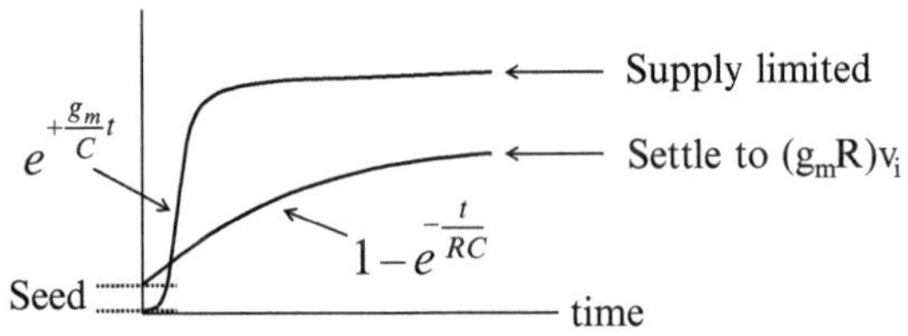

$$H(s) = \frac{\left(\dfrac{g_m R}{1 + sRC}\right)^2}{1 - \left(\dfrac{g_m R}{1 + sRC}\right)^2}, \tag{1.7}$$

which gives a positive real axis pole approximately at $+g_m/C$. The latch pole is $g_m R$ times higher than the negative real $-1/RC$ pole. The positive real pole implies that the latch is unstable, and any transient seed signal will grow exponentially as exp $\{(+g_m/C)t\}$. That is, the latch output grows much faster with a positive exponential term while the preamplifier output settling error decays with an RC time constant as shown in Fig. 1.5. Note that the slope of the transient output of the latch at the latching moment is also steeper than that of the preamplifier by the gain factor of $g_m R$.

The latch can be implemented with and without clock. The latter is a self-latching circuit that derives the seed signal from the input, and the positive feedback is enabled after the seed signal is amplified to be large enough to override the latch hysteresis. However, in most practical uses, the latch needs clock. The clock can initiate the positive feedback of the latch. In most clocked latches, the seed signal is set as an initial condition, and destroyed as the signal starts to grow due to the positive feedback. Both preamplifier and latch repeat the reset and amplification phases as shown in Fig. 1.6.

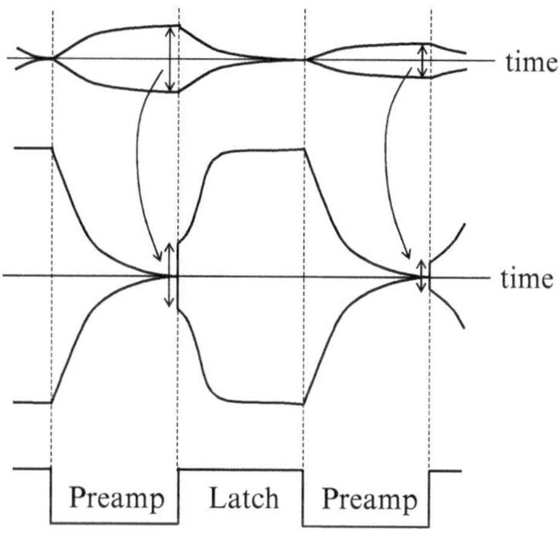

Fig. 1.6 Preamplifier and latch operations

The preamplifier amplifies when the latch is reset while it is reset when latching. The latch input is initialized at the end of the pre-amplification period. However, as the ADC sampling rate gets higher, even the latch cannot complete flipping its output within a given limited time. That is, the latch gain is defined as

$$A_{\text{latch}} = e^{\frac{g_{\text{m}}}{C} \times \frac{1}{2f_{\text{s}}}} = e^{\frac{g_{\text{m}}}{C} \times \frac{T}{2}}. \tag{1.8}$$

Since the polarity of the signal only matters, the latch output needs to change by half the supply voltage within half the clock period. Otherwise, a metastable condition can be reached, yielding invalid no-decision output.

1.1.5 Preamplifier Bandwidth

The function of the preamplifier is to grow a seed signal so that the latch can make decisions with acceptable BER. Thus its gain can be set accordingly, but its bandwidth requirement becomes very stringent depending on whether comparator's input and output are reset or not. If non-resetting preamplifiers are used as in SAR ADCs, they should settle with a full accuracy of $1/2^N$ since the preamplifier output should recover a small seed input of a fraction of an LSB from the previous full-range output during half the clock period. Therefore, the bandwidth of the preamplifier should be wide enough to recover from switching transient and to settle with a time constant of

$$\tau = \frac{1}{\text{BW}} < \frac{1}{2\ln 2^N} \times \frac{1}{f_s}, \tag{1.9}$$

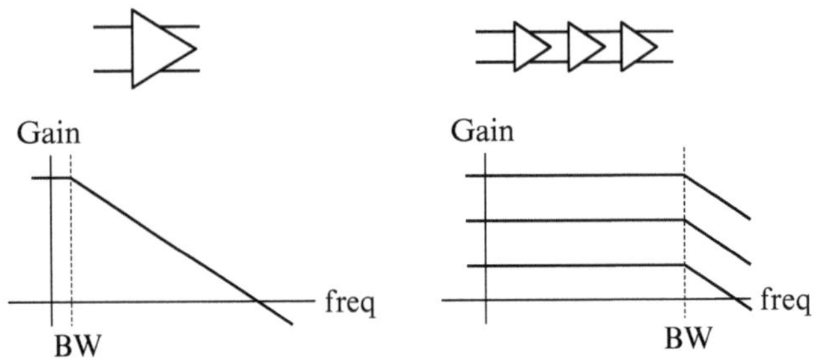

Fig. 1.7 Bode plots of opamp and comparator preamp

where $1/\tau$ is the open-loop bandwidth (BW) of the preamplifier. For 10b ($N = 10$), settling with about 6.9 time constants is required within the given half clock period.

However, if the preamplifier is reset as shown in Fig. 1.6, its open-loop bandwidth can be narrower, and settling with only 3–4 time constants suffices. The logic behind it is that if reset, the preamplifier can settle from the reset condition with only minimum reset errors such as switch feed-through and charge injection. Since only the polarity of the error matters, the resetting preamplifier doesn't need to settle accurately. In spite of that, narrowband amplifiers like compensated opamps cannot be used as comparator preamplifiers because high-impedance output nodes respond to sudden transients very slowly with too long time constants.

The bandwidth requirements for opamps and comparator preamplifiers are compared in Fig. 1.7. Opamps are always frequency-compensated and narrowbanded so that they can stay stable for feedback. On the other hand, preamplifiers are open-loop amplifiers, and should meet both high gain and bandwidth requirements. Since the gain-bandwidth product is constant for any given processes, preamplifiers are implemented by cascading multiple stages with low gains but wide bandwidths.

1.2 Dynamic Amplifier and Latch

Once latched into one of the bistable states, the output of the static latch is almost irreversible. If it is driven hard with a stronger inverter, the output flips but high-level short current flows, which is also translated into large hysteresis of the latch threshold. Therefore, clocked dynamic latches have been commonly used to save high short current and reduce input hysteresis by resetting. Two examples and their latching operations are shown in Figs. 1.8 and 1.9, respectively. The latch on the left side is a standard clocked latch, and the other is a modified version to facilitate the seed initialization and to remove the latch bar clock. Note that the latter conducts current as dotted after latched.

In the standard dynamic latch shown on the left, the differential input seed signal is sampled on the capacitors. When latched, the seed is destroyed, and the output is

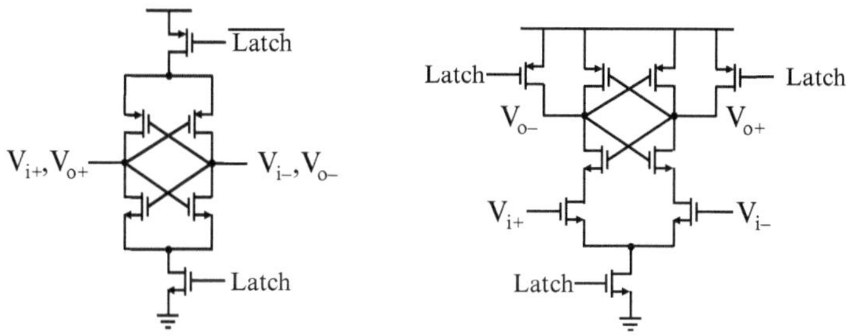

Fig. 1.8 Two dynamic clocked latch examples

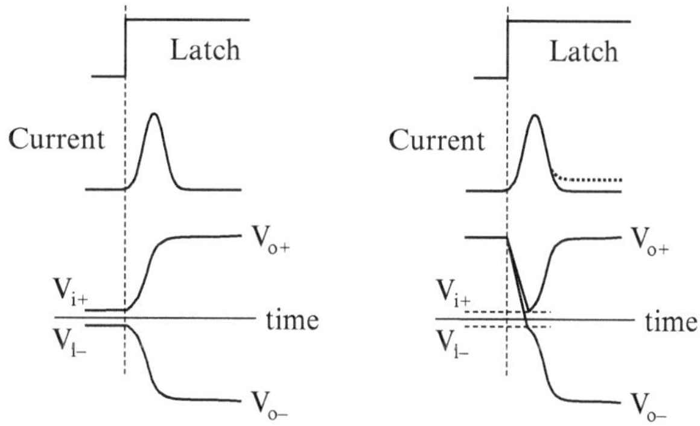

Fig. 1.9 Latching operations of two dynamic latches

split by positive feedback until they hit the supply rails. This latch is common in dynamic random access memory (DRAM). Note that the spike-like short current flows as in the digital inverter during the transition. In the dynamic latch shown on the right, a floating differential pair injects the seed current, and two parallel switches reset the output to the high supply before latching [1].

The common source node of the two input transistors is undefined and floating unless properly reset. The seed input is sampled on the input capacitances that are small since it is normally turned off before latching while both the latch outputs are parked at the high supply voltage. Once latched from this initial condition, both outputs begin to fall together, but one side is pulled down more strongly than the other due to the imbalance of the seed input. The positive feedback mechanism regenerates the seed signal so that it can be split into high or low voltages. To see what is causing the initial slew from the high supply, the latch should be considered as a dynamic amplifier first [2].

A dynamic amplifier shown in Fig. 1.10 works on the same principle as the clocked latch does except for using the static current source load in place of the dynamic positive feedback load. It is made of a differential pair inserted into the inverter, and its tail and bias currents are switched like an inverter. The dynamic amplifier is stable while the latch is unstable since the load resistance is negative. However, their operations are identical during the period the outputs slew down from the high supply voltage. As shown on the right side, the bias current spikes

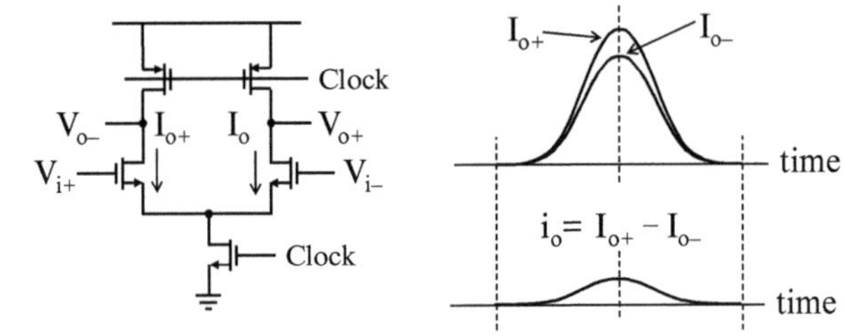

Fig. 1.10 Dynamic amplifier

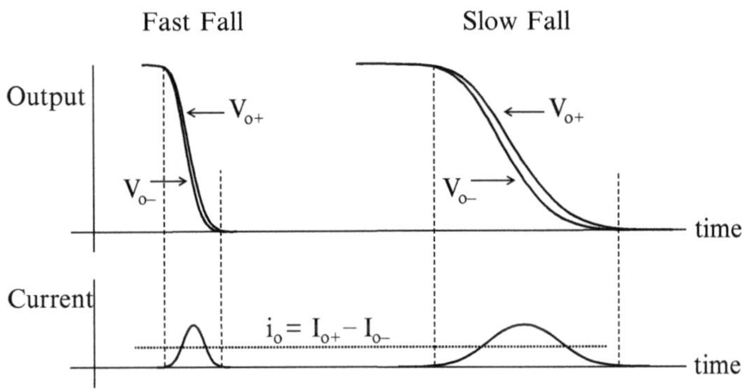

Fig. 1.11 Fast and slow transient outputs

briefly only after the tail current is turned on but until the PMOS bias currents are turned off.

If the clock goes high, both output nodes are pulled down together by two currents I_{o+} and I_{o-}. Since one of the current is higher than the other due to the differential seed input, one output is pulled down a little faster than the other. The small difference ($i_o = I_{o+} - I_{o-}$) between two pull-down currents is also shown. While the output common-mode voltage slews down lower, a differential output voltage develops since two outputs slew with different rates. However, due to its dynamic nature, the amplifier output is active only during the brief period when the bias current flows as sketched in Fig. 1.11, which shows two cases of fast- and slow-fall cases.

The differential output can be estimated from the differential output common-mode voltages as follows.

$$v_o = V_{o+} - V_{o-} = \left(\frac{I_{o+}}{C} - \frac{I_{o-}}{C}\right) \times \Delta t = \frac{i_o}{C} \times \Delta t = \frac{g_m v_i}{C} \times \Delta t, \tag{1.10}$$

where C is the loading capacitor [3]. Although the bias current spikes briefly, the small difference current i_o is assumed to be an average constant current for the linear small-signal transient analysis. The output voltage developed in one transition is the integration of the differential current on the loading capacitor during the transitional period since the integration time for the differential small signal can be estimated.

In general, each circuit node has its own unique time constant given by the product of the total R and C values of the node. Then the node voltage settles exponentially when driven by a high-impedance current source like transistor. If its trans-conductance is g_m, the amplifier has a small-signal gain of g_mR and a bandwidth of $1/RC$. However, if enough settling time is not allowed as in this transient fast-fall case, the node voltage slews at the highest rate of g_m/C, which is also defined as the unity-gain bandwidth ω_T. Consider the highest frequency sinewave of

$$v_o(t) = A \times \sin \omega_T t = A \times \sin \frac{g_m}{C} t. \tag{1.11}$$

It changes with the highest slew rate of ω_T at zero-crossings. Therefore, the steepest slope at zero-crossings is given as a function of the magnitude A and the unity-gain frequency ω_T.

$$\mathrm{SR} = \left.\frac{dv_o(t)}{dt}\right|_{t=o} = A\omega_T = A \times \frac{g_m}{C}. \tag{1.12}$$

Then the slewing time to reach the magnitude A with this slew rate is

$$\Delta t = \frac{A}{A \times \frac{g_m}{C}} = \frac{C}{g_m}. \tag{1.13}$$

This implies that the voltage gain in the fast-fall case is unity since the unity-gain bandwidth is defined as the frequency that the signal can slew up and down at the maximum rate.

$$\frac{v_o}{v_i} = \frac{g_m}{C} \times \Delta t = \frac{g_m}{C} \times \frac{C}{g_m} = 1. \tag{1.14}$$

However, in the slow-fall case, the differential pair responds to the input seed with a time constant RC of the output node. So the small-signal gain of g_mR can be obtained if longer time is allowed for settling. The unity-gain bandwidth of g_m/C is the gain bandwidth product. For the small step input, the output approaches the amplified output exponentially. Therefore, the normalized output is given by

$$\frac{v_o}{v_i} = g_mR \times \left(1 - e^{-\frac{t}{RC}}\right). \tag{1.15}$$

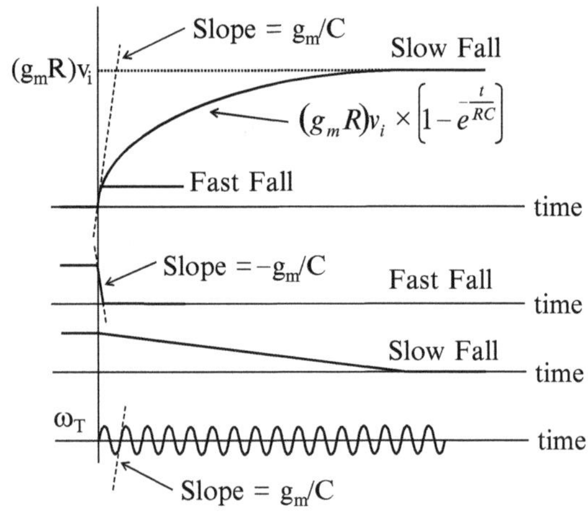

Fig. 1.12 Nonlinear slewing vs. linear settling

By differentiating it, the slope at $t = 0$ can be obtained as g_m/C, which is the slew rate at the beginning as explained in Fig. 1.12.

The slew rate given by the slope at the beginning is the same as the unity-gain frequency ω_T. The fast- and slow-fall cases are also sketched using the same time scale. Note that the waveform with the unity-gain frequency is added in the figure for the illustration purpose. In the fast-fall case, it slews with the same rate of g_m/C, but the gain is just unity. On the other hand, in the slow-fall case, it settles exponentially with a time constant of RC, and has a high gain of $g_m R$. The dynamic latch operates similarly like the dynamic amplifier at the beginning. Two outputs parked at high supply voltages slew down and split, and the positive feedback latch takes over as the common-mode bias voltage falls to activate the latch. The latch outputs are separated and grow exponentially by $\exp(+g_m t/C)$ until they hit the supply rails.

Figure 1.13 illustrates the two cases of amplifier settling and slewing. If the input signal is small, all amplifiers or inverters with high output impedance operate in a linear mode, and the output settles to the final value of $(g_m R)v_i$ exponentially with an RC time constant. However, if the input is switched rail to rail like the digital signal, the output cannot settle to the final value which is too large as the output is limited by the supply. Since all exponential settling starts with a high slew rate of g_m/C, the initial slewing gives a wrong impression that the gain can be higher than is possible with the small-signal settling. Open-loop amplifiers slew and settle. The rise and fall times of digital inverters can be approximated with slew rates, but the linear small-signal gain of analog amplifiers can only be analyzed with settling time constants. That is, all dynamic preamplifiers should stay powered on for longer fine settling. All comparator preamplifiers should be designed to settle within a given time.

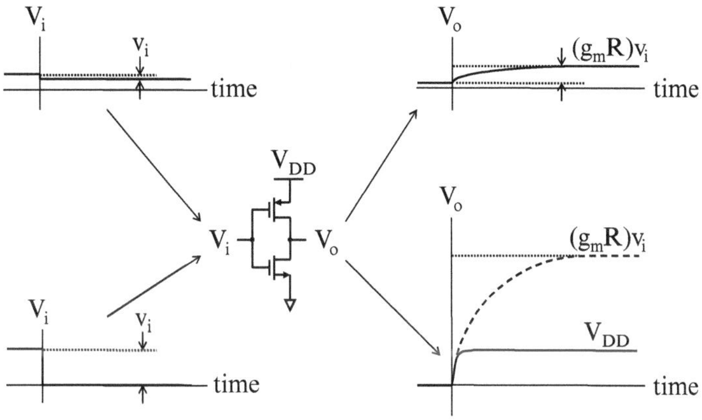

Fig. 1.13 Small-signal settling vs. large-signal slewing

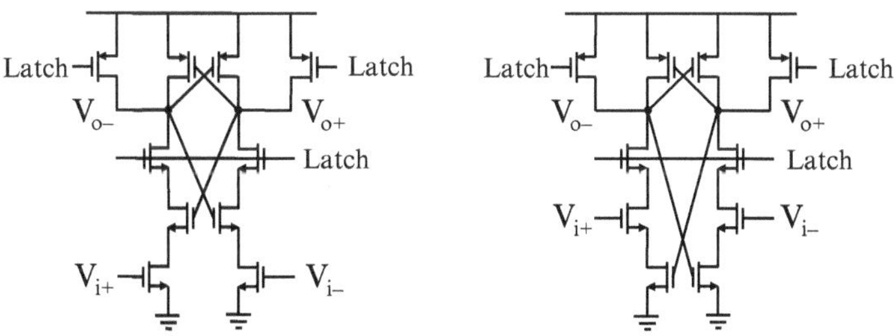

Fig. 1.14 Latches with seeds on grounded and floating inputs

1.2.1 Kickback in Dynamic Latch

The seed input is affected by many factors such as static offset, comparator error, and latch hysteresis. The same is true to the transient noise case. Gaussian white noise is lower in the fast-fall case than in the slow-fall case since the transitional period for the noise to come out is short. As the signal doesn't grow in the fast-fall case either, the SNR stays about the same. However, the transient noise is much higher than in the steady state since the switches are weakly turned on during the transitional period except for the brief period of the peak short current in the middle.

If the bottom latch switch is split into two and inserted inside the latch, the latch is modified as shown on the left side of Fig. 1.14. Note that the input seed signal is now sampled on the grounded transistors that are off but biased in the triode region. This arrangement offers larger input storage capacitance for seed than the previous floating transistor input. However, when it starts to latch, the static offset will be larger in the triode input case than in the saturation input case. Also the critical drawback of these two cases is that power is consumed as the high side draws current when latched. It is because after latched the higher side still holds the seed input and conducts current even though the low side is turned off. This current after latched can be saved by moving the input transistors to inside the latch as shown on the right side. Then this latch becomes purely dynamic as power is consumed only when the latch is turned on and makes transitions [4]. That is, to save power, the seed input transistors should be floated inside the latch. However, the input capacitance becomes small again since the transistors are normally off when the seed is sampled, and susceptible to kickback from the latch output, which also contributes to the latch hysteresis.

There are four possible ways to feed the seed input into the latch as shown in Fig. 1.15. The input nodes are all capacitive, and charged by the preamplifier. In high-impedance seed inputs using transistors, if the two input transistors are turned on, their capacitors get bigger. The input seed signal stored on the small off-capacitances can be lost by the capacitive feedback called kickback, thereby causing more latch hysteresis. It is because the transistor inverts the seed polarity, and the channel capacitance couples the output transient back to the high-impedance input.

This kickback from the latch output gives the input hysteresis as shown in Fig. 1.16. As the latch outputs flip, the input nodes capacitively coupled to the output nodes are affected when the input transistors are fully turned on. As the seed input initiates the

Fig. 1.15 Four possible seed inputs for latch

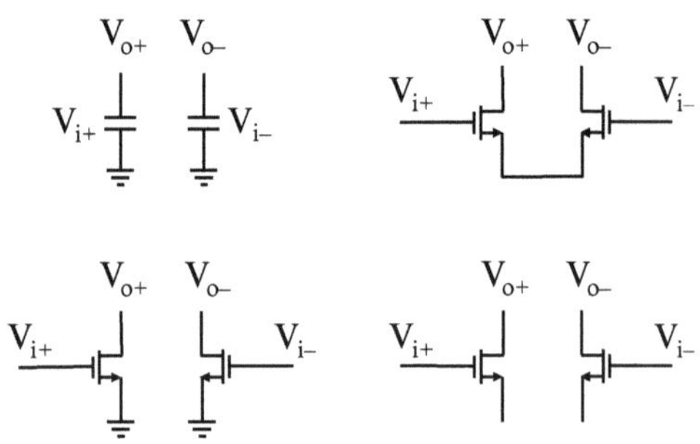

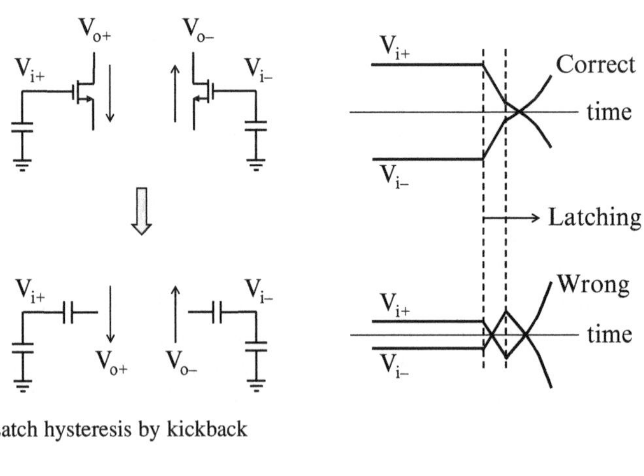

Fig. 1.16 Latch hysteresis by kickback

output latching process, the high input pulls down its output faster than the low input, and the output voltages start to be separated. If the high-input transistor is turned on, it momentarily pulls down the high-input node since the coupling capacitance gets larger suddenly. As the low-input transistor is also turned on, the low-input node is pulled up. Therefore, the polarity of the weak seed input stored on the off capacitance can be reversed due to the capacitive kickback which occurs during the brief transition period marked by two dashed lines. This kickback demands that the input be initialized with larger seed signals than necessary to override the latch hysteresis. It also reduces the seed input by the capacitor ratio due to the Miller effect.

1.2.2 Latch Hysteresis and Metastability

Comparator problems all stem from coarse operation of positive feedback latches. Both the latch hysteresis and metastability give sporadic non-Gaussian errors, and greatly affect SNR and BER. The former is static in nature, but caused by many factors such as threshold shift, kickback, and previous history left in high-impedance latch outputs while the latter mainly results from incomplete latching or too small seed. Unless the previous residues left on all nodes are cleared by reset, floating devices are frozen with the residues from incomplete previous latching, and the previous history is memorized in the high-impedance latch output, thereby affecting the latch threshold.

In all dynamic switching circuits, it is critically important to reset all the input and output nodes properly so that they can give fast and accurate transient responses. The latch hysteresis issue is more critical in applications such as in SAR ADC, which makes many decisions without resetting comparators. The prominent benefit of SAR is that it needs no opamps but only comparators. However, the very benefit turns into a handicap at high clock rates because accurate decisions should be made at much higher rates than other Nyquist-rate ADCs. Furthermore, as the decision time period is shortened, the latch gain gets lower, and the latching gets less complete. That is, SAR is more prone to the latch hysteresis and metastability.

Figure 1.17 illustrates the metastability condition of the latch. Either short latching time or small seed signal can lead to the metastable state which outputs neither digital high nor low. The comparator gain and latching time constant given

Fig. 1.17 Latch metastability

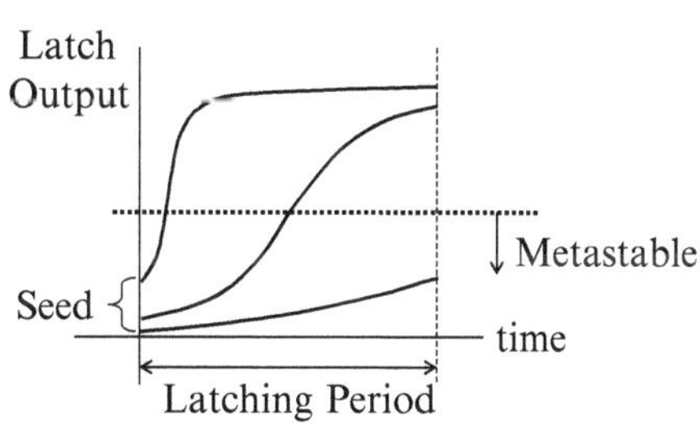

by (1.8) and (1.9) should be warranted for proper latching operation. If the seed input gets small, these errors are inevitable. The common remedy for the hysteresis and metastability issues is to amplify the seed input before latching. The design goal is to lower the probability of getting these errors. Previously, the conditions of 1 dB SNR degradation or 0.1 % BER are estimated using (1.2) and (1.4), respectively. Both conditions give an approximate error bound of 0.15 times the quantization step. Therefore, for this level of error to be amplified to reach the full digital magnitude, the following condition should be met.

$$\frac{0.15}{2^N} \times A_{\text{pre}} \times e^{\frac{g_m}{C} \times \frac{T}{2}} > 1, \tag{1.16}$$

where A_{pre} is the preamplifier gain. The minimum preamplifier gain is therefore

$$A_{\text{pre}} > \frac{2^N}{0.15} \times e^{-\frac{g_m}{C} \times \frac{T}{2}} > 2^N. \tag{1.17}$$

This preamplifier gain requirement is very severe. If 3 and 4 time constants are allowed for latching during *T*/2, the gain requirement is 0.33×2^N and 0.12×2^N, respectively. In practice, most ADC designers may use a rule of thumb number of 2^N. The gain and bandwidth of preamplifiers are the most critical ADC design parameters. They should be warranted to be high and wide enough for proper ADC operation with negligible hysteresis and metastability errors.

1.2.3 Transient Noise in Switching Circuits

The white noise power spectral density is defined as an average power per unit bandwidth, and should be handled in the frequency domain. It is an AC steady-state parameter assuming that the circuit environment such as biasing is stationary and time-invariant. However, like the sampled wideband noise, switching analog circuits such as dynamic amplifiers, latches, mixers, and voltage-controlled oscillators (VCOs) operate in transients and even time-varying modes, and the noise performance of such switching circuits is of prime interest.

The noise characteristic of an inverter is illustrated in Fig. 1.18. Since noise is defined as the power spectral density in steady state, it is not correct to visualize any time-domain instantaneous voltage waveform. What is conceptually sketched is the fluctuation of the instantaneous power in the time domain. Even this concept of the instantaneous power is based on the fact that the circuit is in steady state with a fixed bias condition kept constant. As the input goes high, the short current flows after the NMOS is turned on, but is cut off when the PMOS is turned off. That is, noise fluctuates as the bias condition changes.

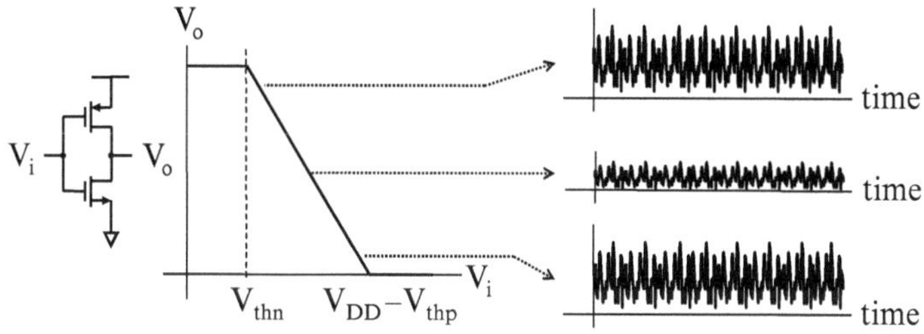

Fig. 1.18 Transient noise characteristics

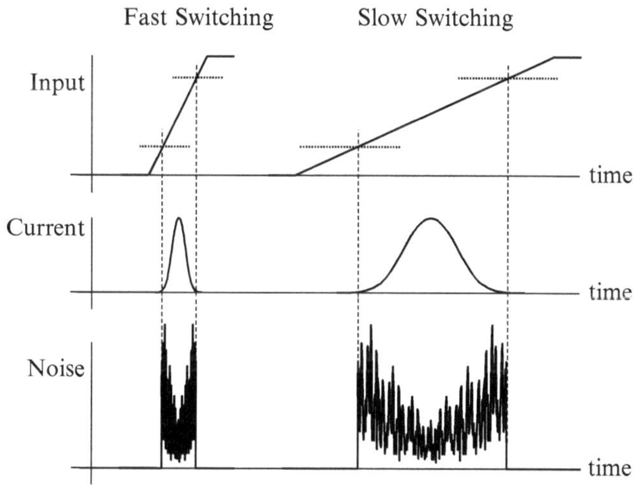

Fig. 1.19 Conceptual explanation of the transient noise

Although it is not possible to analyze the transient noise of switching circuits in the time domain as in this inverter example, the instantaneous power can be assumed to fluctuate over the extended time period. The transient noise strongly depends on the slope of the rise and fall transitions of the switching circuit. When transistors are turned off, they have no noise. In the fast switching case, the period the noise comes out is shorter than in the slow switching case. Since the noise power in the time domain can be obtained using the fast Fourier transform (FFT) of its autocorrelation function, noise will be lower in the former case than in the latter case. Therefore, making fast switching transitions has been the most common practice in the designs of switching circuits such as low-noise mixers, ring oscillators, and VCOs.

In Fig. 1.19, the equivalent transient input noise is conceptually explained for the fast- and slow-fall cases assuming the input changes slowly. Note again that this figure is somewhat artificial since noise during the brief switching transient period cannot be defined. Noise power can be defined only when the bias condition stays constant for an extended period. Assuming that the bias condition is constant at any

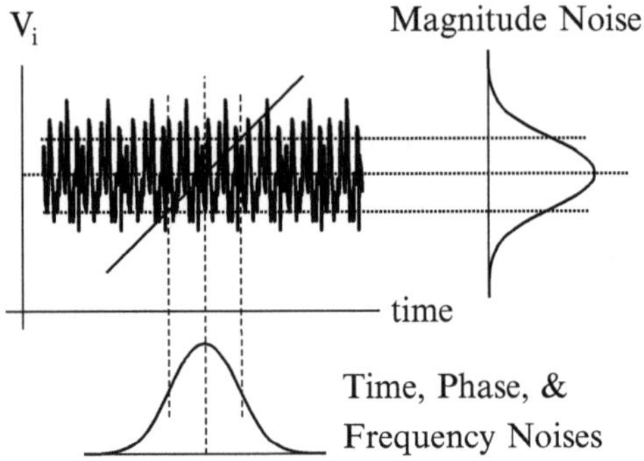

Fig. 1.20 All noises are related

transient states, the equivalent input thermal noise spectral density of a switching inverter biased with current I is approximately

$$\frac{v_{\mathrm{n}}^2}{\Delta f} = 4kT \times \left\{ \frac{2}{3\left(g_{\mathrm{mn}} + g_{\mathrm{mp}}\right)} \right\} \propto \frac{1}{\sqrt{I}}. \tag{1.18}$$

The current peaks at the middle of the supply, but the input noise peaks at both ends since it is inversely proportional to the square root of the current. That is, switching circuits are very noisy when transistors are weakly turned on during the transitional period. If turned off, no noise comes out, and if fully turned on at the middle of the supply where the bias current peaks, noise is low.

That is, the transient noise of switching circuits is difficult to quantify as it varies widely with the bias condition and also with the rise/fall times of switching transients. In particular, the noise performance of the dynamic amplifier and latch is severely affected by the slope since they operate in open loop. Even in the closed-loop VCO design, only the noise at zero-crossings matters since oscillation grows as the small signal grows. What affects the VCO noise most is the quality factor Q of the tuning circuit, which helps to reduce noise variance by filtering at higher offset frequencies. That is, high-Q tuning limits the bandwidth of the noise spectrum, thereby reducing the in-band noise power. The lower the magnitude noise gets, the lower the other time, frequency, and phase noises become.

All noises in magnitude, time, frequency, and phase are related by the slope of transition as shown in Fig. 1.20. This time-domain noise is just a conceptual sketch. Let's consider the dynamic amplifier as shown in Fig. 1.10. The differential output is obtained by integrating two different current spikes on the output capacitances while the differential pair is turned on and conducts current. If the current spike is assumed to be constant over the integration period Δt, the standard way to estimate the integrated noise variance is to use a gate function in the time domain for moving average filtering. It is like sampling the transient noise power only during Δt. In the frequency domain, it is equivalent to filtering with a lowpass function of $\mathrm{sinc}(\pi f \Delta t)$, whose equivalent noise bandwidth is $1/2\Delta t$. Therefore, the equivalent input noise of the dynamic differential amplifier is estimated as

$$v_{\mathrm{n}}^2 = 2 \times 4kT \times \frac{2}{3g_{\mathrm{m}}} \times \frac{1}{2\Delta t} = 2 \times 4kT \times \frac{2}{3g_{\mathrm{m}}} \times \frac{g_{\mathrm{m}}}{2C} = \frac{8}{3} \times \frac{kT}{C}, \tag{1.19}$$

where the integration time Δt can be assumed to be the slewing time for the fast fall case. Here it is also assumed that the initial noise before integration is independent of the noise to be integrated. It is about the same as the kT/C noise.

1.3 Analog Integrate and Dump

In the slow-fall case, the dynamic amplifier operates like a normal differential pair with the same noise characteristic. Thus its output settles exponentially with an *RC* time constant. The equivalent input noise decreases as the equivalent noise bandwidth gets narrower since the slewing time is no longer limited. However, if long enough settling time is not allowed in the fast-fall case, the dynamic amplifier operates like an integrator. The integrate-and-dump concept is widely used in digital signal processing. It is a time-domain matched filter used to detect the impulse symbol spread over the symbol period and to average the wideband white noise. However, it requires proper caution to use it for accurate analog signal processing.

Analog integrator is DC unstable, and as a result, it is seldom used as an open-loop amplifier. They are mostly used to mimic energy storing elements in analog filters which are DC-biased and stabilized by feedback. Integrator gain is determined by the product of the slew rate and the integration time. Neither parameter is well defined. Since the current changes rapidly in this dynamic biasing situation, the integration slope becomes very nonlinear. Therefore, it is not possible to implement accurate voltage gain stages using integrators. However, coarse gain stages for comparators can be made dynamically using integrators assuming slewing starts from correct initial condition. Although the white thermal noise can be averaged out by integration, the signal is also shaped by the same integration slope as explained in Fig. 1.21.

In fact, integrators are just low-pass filters with very low-corner frequencies. Unless the signal spectrum is concentrated at almost DC while noise is white, integrators have no definite advantage in improving SNR since the signal spectrum is also tilted by the same integrator slope. However, in reality, SNR gets even worse. It is because the integrator pole is at DC or at very low frequency, and the integrator responds to low-frequency input very slowly. The integrator output drifts away since the initial condition is set by DC errors including offset and $1/f$ noise. Thus integrator should be reset to an exact initial DC condition before it starts to integrate.

The integrator's unity-gain bandwidth determines the integrated output level. Slow integrators integrate less noise, but the integrated signal is also small. On the other hand, fast integrators integrate more noise, but the integrated signal is also large. Integration time is working as a gating function both for the signal and noise. Fast integrators have two drawbacks: DC instability and difficulty in turning bias

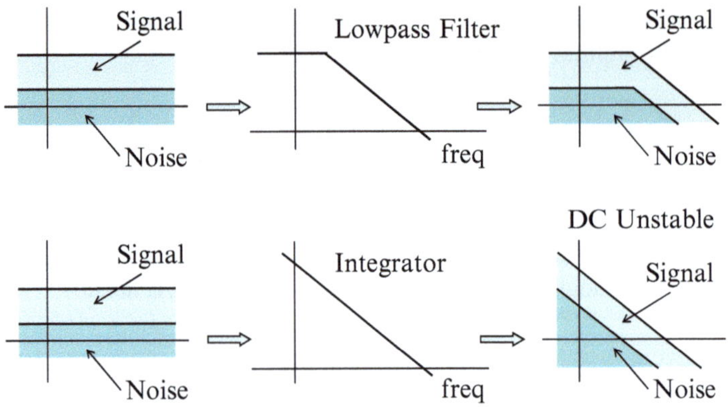

Fig. 1.21 Lowpass filter vs. integrator

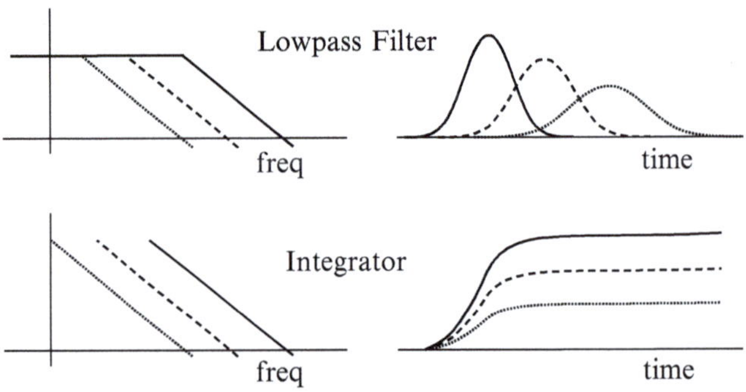

Fig. 1.22 Frequency and time responses of the lowpass filter and integrator

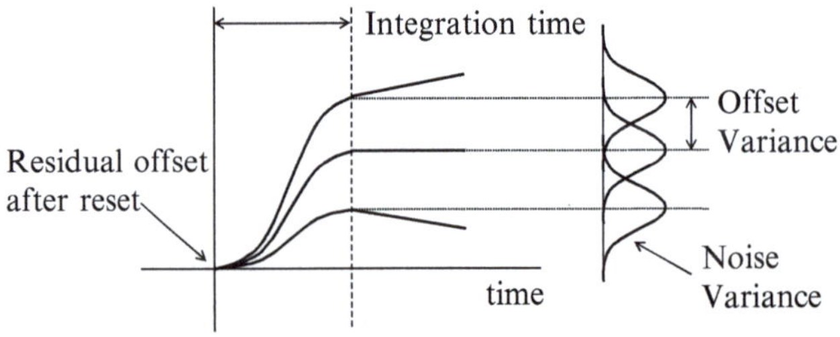

Fig. 1.23 DC instability of the integrator output

on/off. Analog switched-capacitor integrators in any forms have residual offsets after reset due to the switch feed-through and charge injection. Conceptually, lowpass filtering spreads the impulse over time while integrating limits the integrated output magnitude as shown in Fig. 1.22.

The integrator output is undefined due to the DC instability as explained in Fig. 1.23. Integrator has both constant and dynamic offsets. If the offset variance is taken into account, integrator exhibits poor SNR. Integrators often find applications that need to convert voltage into time as in the slope-type ADCs or time-to-digital converters [5]. It would be challenging to design a simple but fast analog integrator with negligible reset offset error. A practical integrator example is a coarse PLL charge-pump circuit.

1.4 Quantization Error vs. White Noise

The ADC output in time domain can be divided into two. One is deterministic and the other is random. The magnitudes of the deterministic terms can be added.

$$S_o = \sqrt{S_i} \pm \sqrt{Q} \pm \sqrt{D} \pm \delta_c, \tag{1.20}$$

where S_i, Q, and D represent the powers of the input, quantization error, and distortion, respectively. The comparator error term δ_c in a broad sense includes all known non-Gaussian comparator-related errors: Threshold, sparkle, and metastability errors. In (1.20), the quantization error is a random white noise, and its spectrum is assumed to be white. The non-Gaussian portion of the quantization error can be separated and defined as distortion. The distortion and comparator threshold errors are two deterministic error portions in the ADC output. However, the static distortion error is the only error that can be reduced by trimming or self-calibration techniques as it stays constant once the DC transfer function of the ADC is fixed. Although it is the same deterministic error, the comparator threshold error cannot be calibrated since it is induced by the undefined hysteresis and metastability.

The total instantaneous output power is the power sum of all including the noise N and jitter J powers.

$$S_o^2 = \left(\sqrt{S_i} \pm \sqrt{Q} \pm \sqrt{D} \pm \delta_c\right)^2 + N + J = S_i + Q + D + \delta_c^2 + N + J. \tag{1.21}$$

Since they are all uncorrelated, their cross-product terms are averaged to be zero over time. The sign and magnitude of the white Gaussian noise and jitter are not defined either. For those random variables, only power matters. Therefore, the noise power spectral density is modified by filtering as shown in Fig. 1.24.

Since the noise is a random variable with a zero mean, it can be defined only using its variance. As in Fig. 1.25, the noise voltages and their algebraic sum and difference in the time domain cannot be defined, but the noise power is defined. The instantaneous power fluctuates forever with a variance assuming that the circuit stays in steady state keeping the same small-signal parameters. However, in

Fig. 1.24 Noise power shaped by filter

Fig. 1.25 Noise power in time domain

v_1	1	−1	1	−1	Wrong
v_2	−1	1	−1	1	Wrong
$(v_1 + v_2)^2$	0	0	0	0	Wrong
$(v_1 - v_2)^2$	4	4	4	4	Wrong
$v_1^2 + v_2^2$	2	2	2	2	Correct

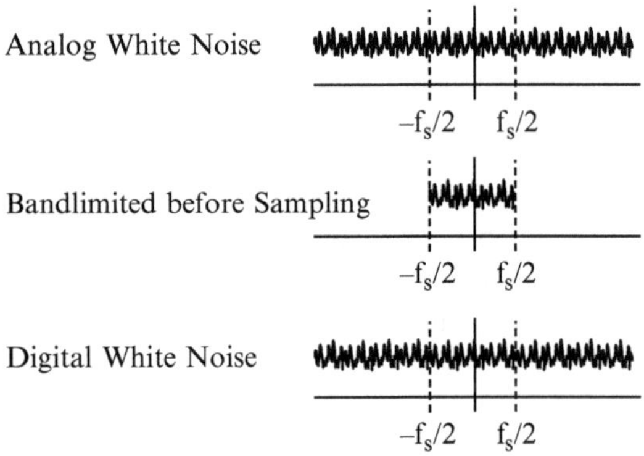

Fig. 1.26 White noise power spectral density

transient circuits, the parameters are changing in real time. Therefore, the transient noise simulations of switching devices such as mixers and VCOs in the time domain are not valid though it may provide good approximate values for analysis purpose.

The same thing is true to the generation of the white noise spectrum in the frequency domain as shown in Fig. 1.26. In principle, the digital white noise spectrum generated using a random number generator cannot represent the true sampled analog white noise spectrum band-limited within the Nyquist band since it is not feasible to band-limit the real white noise spectral density only up to the Nyquist band. Therefore, the sampling process inevitably aliases the unfiltered high-frequency white noise into the Nyquist band. However, generating a white noise spectrum digitally is considered adequate for practical simulation purposes.

1.5 Noise Implications of Sampling

Time-domain analog signal processing relies on the time delay as in digital signal processing, and analog switch is a basic element in sampled-data analog processing circuits. First, switching circuits are nonlinear and noisy. Ideal analog sampling is freezing analog voltage in time. It is a challenging task requiring utmost accuracy both in magnitude and in time. The track and hold mechanism can be implemented using a switch and a capacitor. The voltage on the capacitor at the time when the switch is turned off is sampled on the capacitor. This in effect is to multiply the input by the unit impulse sequence in time. That is, unless the sampling network is band-limited, the noise spectrum spread over the infinite bandwidth is aliased into the Nyquist baseband, and the sampled noise spectral density would rise to be infinite. In the frequency domain, it is to convolute the input spectrum with the impulse spectrum repeated at every f_s as shown in Fig. 1.27, where the input voltage source represents the noise source.

In practice, the finite resistance of the sampling switch also gives many implications on its noise and circuit performance as shown in Fig. 1.28, where the switch is replaced by the finite resistance R_{on} and its own noise source.

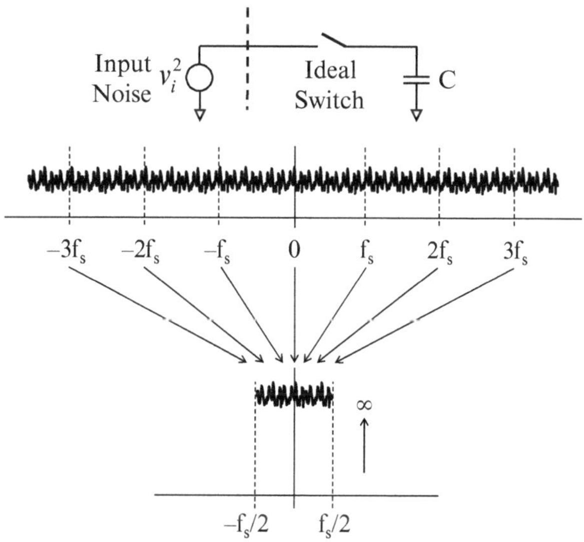

Fig. 1.27 Noise aliased into the Nyquist band

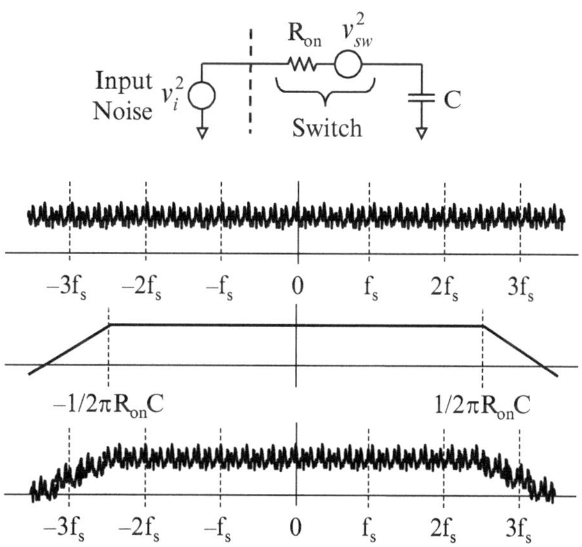

Fig. 1.28 Band-limited white noise is sampled

The switch implemented using NMOS or CMOS transistor is not ideal at all. It has a finite resistance modulated by the gate voltage, and holds a channel charge nonlinearly proportional to the overdrive voltage. Furthermore, it is loaded by junction diodes on both sides, and has overlap capacitances. It is well known that the charge injection and the clock feedthrough leave the nonlinear DC error voltage on the capacitor when it is turned off. Analog switching also contributes to the transient distortion of the switching circuits. Sampled-data circuits are configured so that their effects can be minimized and not affect the performance. Various techniques have been developed such as bottom-plate sampling, offset cancellation, and correlated double sampling (CDS).

The wideband noise is band-limited by the sampling bandwidth of $1/2\pi R_{\mathrm{on}}C$ before it is sampled on C. However, the switch on-resistance R_{on} should be minimized so that the sampling bandwidth can be broad-banded to ensure accurate sampling to meet the resolution requirement. As the sampling error decays exponentially, to sample the input on C with N-b accuracy, the following condition should be met.

$$e^{-\frac{1}{2R_{\mathrm{on}}Cf_{\mathrm{s}}}} = \frac{1}{2^N}. \tag{1.22}$$

Since the noise spectrum rolls off with one pole, the actual sampling bandwidth should be scaled by the factor of $\pi/2$. Therefore, the minimum sampling bandwidth is given by

$$\frac{1}{2\pi R_{\mathrm{on}}C} \times \frac{\pi}{2} = \ln 2^N \times \frac{f_{\mathrm{s}}}{2} \approx 8.3 \times \frac{f_{\mathrm{s}}}{2}, \tag{1.23}$$

for 12b accurate sampling ($N = 12$). This implies that the thermal noise of the source side within the bandwidth 8.3-times wider than the Nyquist bandwidth ($f_{\mathrm{s}}/2$) is sampled, which increases the sampled thermal noise by 9.2 dB.

If the input noise is ignored, only the white thermal noise of the switch is sampled, which gives the well-known sampled kT/C noise which is independent of R_{on}.

$$4kTR_{\mathrm{on}} \times \frac{1}{2\pi R_{\mathrm{on}}C} \times \frac{\pi}{2} = \frac{kT}{C}. \tag{1.24}$$

Note that both the noises from the input source and the switch resistance are band-limited by the finite bandwidth of the sampling network. This sampled wideband noise has been limiting the dynamic range of all switched-capacitor circuits.

Switched-capacitor circuits use opamps to accurately transfer charges (voltages) from one capacitor node to the next one. Opamp designs vary widely depending on the goals to achieve. Most opamps used for switched-capacitor applications are

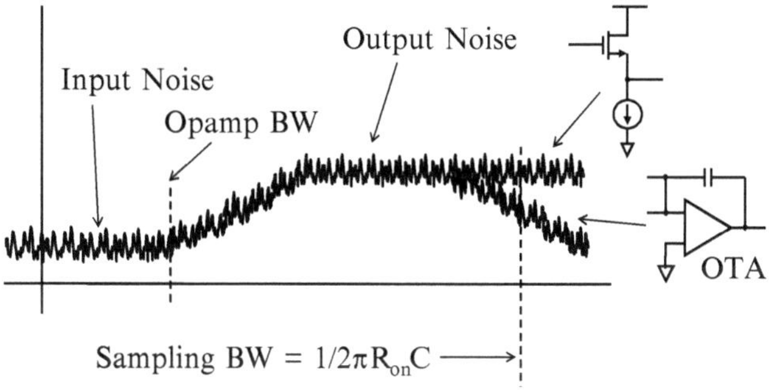

Fig. 1.29 Opamp wideband output noise spectrum

output trans-conductance (OTA) type opamps with high output impedance as it only needs to drive capacitive loads. However, for unity-gain buffers such as in Sallen-Key filters, a simple source follower with low output impedance is also used to save power.

Figure 1.29 illustrates the output noise of the opamp stage conceptually. In the two-stage OTA which has been commonly used in switched-capacitor circuits, the output noise is dominated by the noise of the opamp input stage due to the high DC gain. However, as frequencies go higher, the opamp gain starts to drop, and the output noise rises until the noise from the output stage dominates. If the opamp is made using OTA, the output noise decreases after the unity loop-gain frequency. On the other hand, if the low-impedance output buffer is used, the output noise stays high up to the wideband output pole frequency. Therefore, if the output noise is sampled using a wideband sampling circuit, the sampled noise will be high particularly in the low output impedance case.

For the reason, switched-capacitor filter has been mostly implemented using OTAs, but still exhibit the higher sampled wideband noise than the continuous-time version. The penalty to pay for wideband sampling amounted to about 15–20 dB, and limited the dynamic range (DR) of switched-capacitor filters to the 50–60 dB range. In the case of oversampling $\Delta\Sigma$ modulators for high-resolution audio, it is possible to attain higher DR since the wideband sampled noise is lower since the white noise is spread over the oversampling bandwidth, but for extremely high resolution above 18b, even the sampling bandwidth has been further reduced until the condition of (1.23) is reached by inserting an extra series resistance at the sampling input to narrow the excess sampling bandwidth. For this reason, switched-capacitor filters helped to digitize the telephone voice band in early 1980s, but quickly gave way to digital oversampling techniques.

1.6 Jitter and Transient Distortion

Analog sampled-data processing is equivalent to digital signal processing as signal is represented by sampled voltages at discrete times. However, unlike digital processing, analog voltages cannot make instantaneous voltage jumps from one sampled point to another. The most desirable transition between sampled voltages is the sample-and-hold (*S/H*) waveform. Two errors in time and magnitude matter. One is jitter, and the other is transient distortion.

Jitter is the timing error when the transition is made as shown in Fig. 1.30, where the 1b data within the symbol period T is clocked with jittery clock. First, the 2-level signal has no static magnitude error, but dynamic errors arise due to the jitter. Two jitter effects are notable. One is the pulse position jitter, and the other is the pulse width jitter. The former raises the random jitter noise floor since correct data come out randomly at wrong times. The latter contributes to more detrimental distortion since the data magnitude is modulated by jitter. However, the error from nonlinearity is spread like the white noise, yielding the SNR of

$$\text{SNR} = 20\log\frac{\Delta t}{T}. \tag{1.25}$$

The phase noise of 1° at 1 GHz is equivalent to an RMS jitter of 2.8 ps, which will lead to 51 dB SNR.

The pulse width jitter effect is the worst in the 2-level case (1b DAC) while there is no magnitude error. On the other hand, the pulse width jitter effect is alleviated in the multi-level DAC case since the jitter magnitude gets smaller, but the static magnitude error degrades the performance. For example, for 2-level (1b) data, the dynamic error due to the pulse width jitter may be dominant while for 16-level (4b) data, the static error resulting from the inaccurate multiple levels may matter more. Other than careful design and implementation, there are no known cures to eliminate dynamic jitter-related errors. The clock chain from the source can be simplified to avoid the jitter accumulation, or PLLs for clock generation or synthesis can be broad-banded for low jitter clocks. However, static level-dependent magnitude errors can be trimmed, self-calibrated, or randomized.

There is a far more important and often neglected dynamic error to consider. It is the transient distortion error. Although jitter is just the timing error, the transient

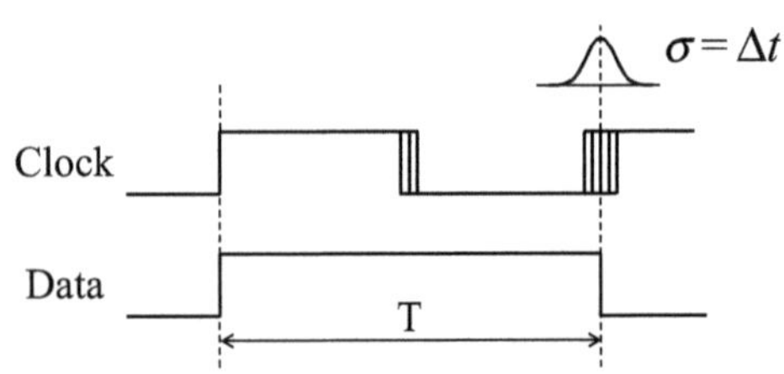

Fig. 1.30 Data timed with jittery clock

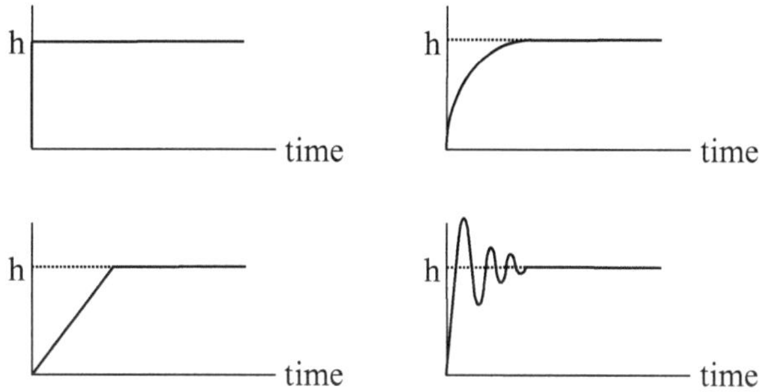

Fig. 1.31 Transient distortions

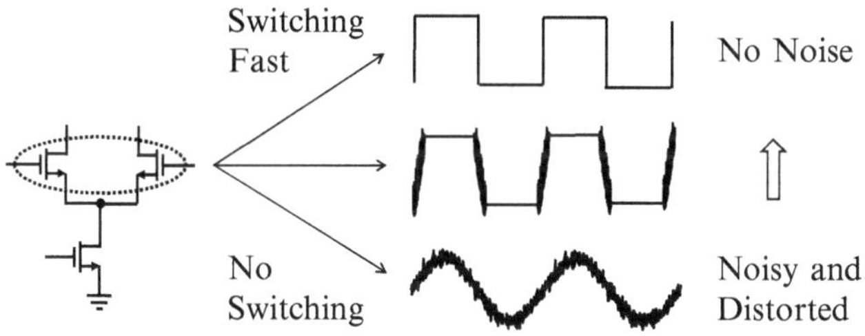

Fig. 1.32 Noise and transient distortion of the chopping mixer

distortion error depends on how analog voltages change from one sample point to another. The only remedy to reduce the dynamic transient distortion is either to make the transition infinitely fast or to control the rising and falling transition with a single time constant.

Figure 1.31 illustrates the four transitions of analog voltages from one sample point to the next. The two cases on the top are linear. One is the ideal *S*/*H* waveform, and the other is an exponential settling with a single time constant. The latter case is linear since the settling error is represented by the area between the ideal step and the exponential settling curve. It is linearly proportional to the height *h* in this exponential settling case. However, in the bottom two cases of slewing and ringing, the area is not linearly related to the height. This nonlinear settling error shows up as transient distortion, and it gets worse at high clock rates. Linearity improves only if either the step is small or the slope is steep. Furthermore, *S*/*H* and analog processing errors are added to the static error, and all switch-related artifacts such as switch nonlinearity, feedthrough and charge injection, and sampled wideband noise get worse at high clock rates [6, 7].

The transient noise and distortion of the chopping mixer are illustrated in Fig. 1.32, where the time-domain noise is visualized only for the illustration purpose. The transient noise of the inverter was explained using Fig. 1.18. The mixer is noisy and distorted when both transistors are turned on. If the chopper steers the tail current by switching the differential pair fast, the time period for the switching transistor noise to come out is shortened, thereby achieving low noise and distortion. The chopper mixer performance depends on the clock rise/fall times.

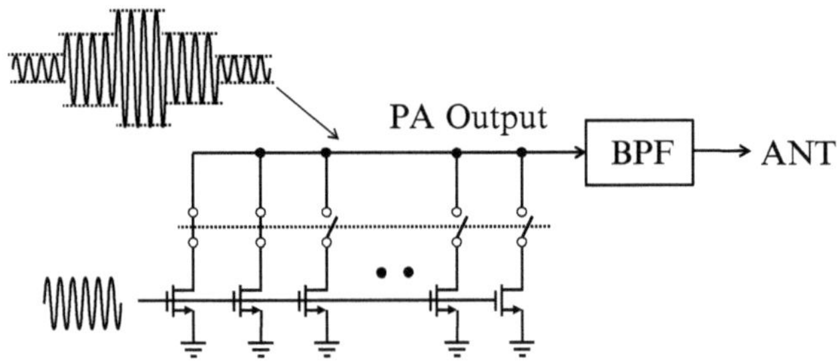

Fig. 1.33 RF TX DAC for power amplifier

The chopper mixer is the weakest element in the RF receiver chain in terms of the noise and nonlinearity performance. Therefore, the RF receiver design hinges on the mixer design. The chopper mixer performs best when it is driven hard. That is, modern RF systems' performance heavily relies on the SFDR of the down-conversion chopper mixer.

Another example of the RF transient distortion is the TX DAC as shown in Fig. 1.33. As switching-based digital RF draws attention, the trend is to generate the RF spectrum directly by modulating constant-envelope signals. It is the classic analog multiplier that changes the envelope, and an utmost care should be taken to warrant transient linearity. Although the pulse waveform is Gaussian-shaped for smooth envelopes, the RF bandpass filter or antenna provides complex poles at the output, which cause poor transient linearity. The transient distortion gives rise to the spectral regrowth and out-of-band emission. Furthermore, analog multipliers, two-quadrant or four-quadrant, are not linear for large PA outputs. Such switching at RF may work with broadband instruments with 50 Ω termination, but in practice, it is considered quite challenging to modulate analog envelope of the RF signal.

1.7 DC Wandering

Handling DC has been the perennial but puzzling issue to analog designers since the era of the old vacuum tube voltmeter. In particular, MOS is a surface device, and its threshold voltage is poorly defined, and even its gate leaks. There are two sources of DC or offset voltages in analog circuits. One is from the source side, and the other is from the analog processing unit such as amplifier. For the latter errors such as opamp offset and $1/f$ noise, many circuit techniques have been developed. The chopper stabilization technique is to modulate the DC or low-frequency component up to high frequencies, and in switched-capacitor circuits, DC or offset voltages can be sampled on capacitors in every clock cycle. However, the CDS is the most general way to avoid the DC offset problem. It is a difference sampling technique at the system level defining the signal as the difference in two consecutive samples. As in all offset cancellations, the penalty to pay is the random noise power is doubled.

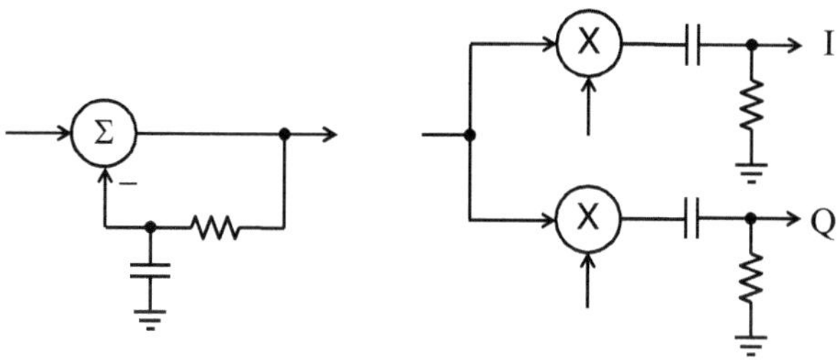

Fig. 1.34 High-pass filtering to remove DC

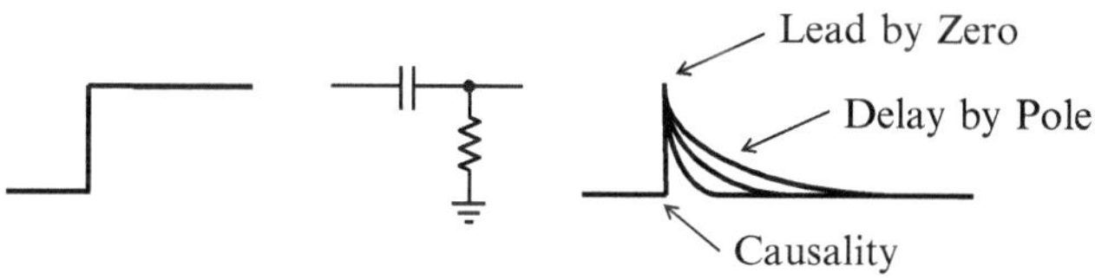

Fig. 1.35 Phase leading transient by zero

It is troublesome in the former case if DC or offset voltages originate from the source side. In particular, if the signal is at almost DC such as in direct-conversion receivers and medical human body sensors, it is almost impossible to separate them from the desired signal in the analog domain. Almost DC signals have two components: Constant DC and time-varying extremely low frequency AC. The common techniques proposed to date are high-pass filtering (AC coupling), DC servo feedback, transformer coupling, and CDS.

Figure 1.34 shows the examples of high-pass filtering or AC coupling. On the left side, the low-pass feedback of the offset is equivalent to high-pass filtering. On the right side, the RF direct-conversion receiver is shown [8]. In-phase (I) and quadrature (Q) mixers down-convert RF spectrum to DC in a complex form of $I + jQ$. The DC offsets in I and Q paths are high-pass filtered. However, the direct-conversion receiver cannot quickly recover from sudden DC offset changes created by power envelope fluctuation and nonlinearity. AC coupling makes DC wander as the low-frequency spectrum is eliminated, thereby causing inter-symbol interference (ISI) in data communications. It works only for steady-state systems with no DC components present.

The DC wandering mechanism in the time domain is explained in Fig. 1.35, and the transient behavior of the zero in the transfer function is illustrated in Fig. 1.36. It always comes with a pole. Usually, phase lags by pole, but phase leads by zero. The phase lead by zero is the response of differentiation in time. Due to causality, the phase lead in the time domain is the phase lead ahead of the pole response. However, in the frequency domain, the actual phase of zero is leading in the steady-state response. Therefore, AC coupling systems only work for stationary inputs with constant envelopes and magnitudes such as from test instruments.

AC coupling circuits with long time constants have undefined DC bias conditions like the high-impedance probe input of voltmeters. The input node DC voltage

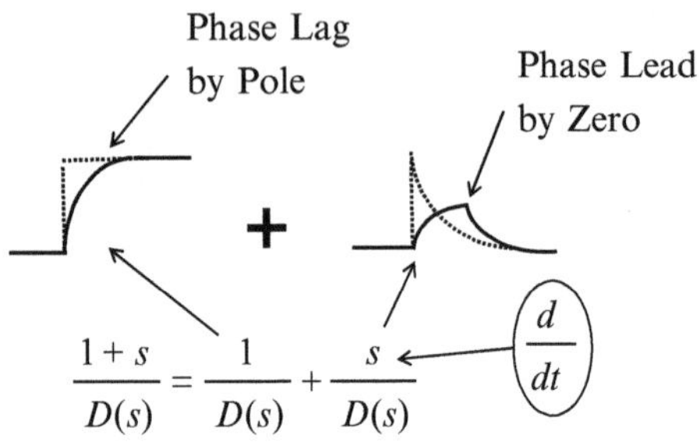

Fig. 1.36 Transient response with a zero

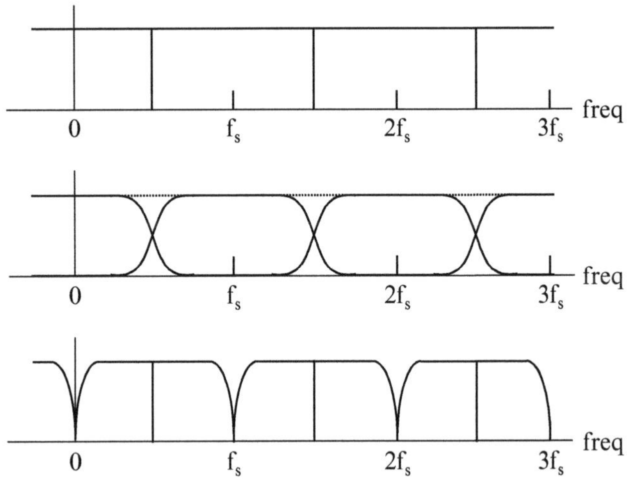

Fig. 1.37 Nyquist-rate spectrum: Ideal, with excess bandwidth, and DC-notched

is very sensitive to the coupling through the input capacitor. In human body sensors like ECG and EKG, such sensitivity is called motion artifact. As the sensor capacitor to body gap varies when body moves, the charge relation of $Q = CV$ makes the voltage change to keep the constant charge. Therefore, if there is a sudden jump in the input DC as shown, its high-pass filtered output is delayed by the long time constant and settles back to a stable DC state. If the input is disturbed, this slow-pole node constantly strives to settle, and creates an undefined DC wandering situation.

This DC wander effect can be explained using the spectrum as shown in Fig. 1.37. In the ideal Nyquist-rate sampled spectrum at the top, signal can be reconstructed without loss. This implies that if any portion of the spectrum is lost, the original signal is lost. However, in most symbol-rate sampling system, the ideal brick-wall Nyquist filter for antialiasing does not exist. Thus the alternative is the vestigial signal band with excess bandwidth. If the power envelope is constant as shown in the middle, there is at least no ISI occurring in the decision. However, the DC-notched spectrum at the bottom violates the constant power envelope rule due

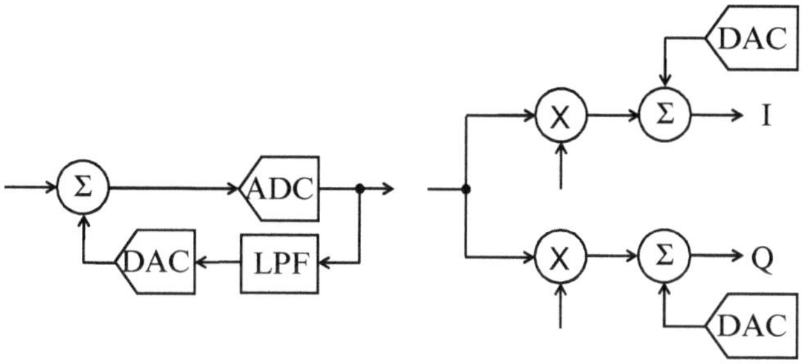

Fig. 1.38 Practical discrete-time DC notching system using digital CDS

to the repeated missing portions of the spectrum. Even though a narrowband high-pass filter is used for AC coupling, the notch repeats infinite times, and it is unlikely for the original signal to be reconstructed. That is, the DC wander makes the signal baseline wander slowly, and significantly affects the BER in the digital demodulation.

At the system level, the only practical DC notching system is the digital discrete-time CDS as shown in Fig. 1.38. To avoid the DC wander in direct-conversion RF receivers, low IF is preferred to DC IF in narrowband or continuous frequency-modulation (FM) systems such as cellphones and Bluetooth. However, for wide-band phase-modulation (PM) systems, the discrete-time feedback using DAC overcomes the shortcomings of the continuous-time low-pass feedback. The basic idea is to update and hold DC or offset values constant during the period that covers one data packet by servo feedback in digital communications. Such CDS has been a staple in signal processing for CCD and image sensor. One horizontal scanned line is referenced to the constant black level, which is to define signal relative to a constant black level.

The DC wander or motion artifact is the most critical design factor in human body sensors for health since the signal spectrum is concentrated almost at DC well below 100 Hz. It needs to detect a mV-level weak signal riding on 100's mV of CM voltages. The signal of about 1 Hz is band-limited with a 0.5 mHz high-pass and a 100 Hz low-pass filters. This low-corner high-pass filter makes its DC wander and drift away constantly. The recent trend is to correct DC and to perform AGC function even in the digital domain. For the purpose, a much wider DR ADC of 16b or higher is used with extra 6–7b (40 dB) DR set aside just to cover large CM inputs.

1.8 Switching for Low Power

One power-saving measure drawing attention these days is operating analog circuits in intermittent transient modes. However, extreme care should be taken in turning any power or bias circuits on and off. DC or high-Q circuits have long time constants, and it takes a long time for any switching transients to die out. The idea of biasing transistors only when they are used prompted some dynamic biasing like the rise/fall transitions of digital circuits. Any transistors used for analog functions should be operated with stable bias condition. Otherwise, the small-signal steady-state analog theory is invalid as small-signal parameters are undefined or time-varying.

In digital signal processing, the integrate-and-dump function is the most important functional block. In analog circuits, it is an integrator which is reset at the beginning and sampled at the end. Therefore, turning the integrator biasing on during the integration cycle is an attractive option for low power. However, an analog integrator has a memory, and operates in open-loop condition, and integrator is DC-unstable. DC offset and low-frequency $1/f$ noise are integrated. For power supply insensitivity, all bias lines should be heavily bypassed to the supplies. That is, bias lines are stabilized long after the bias circuit is turned on. Integrating with time-varying parameters leads to inaccurate outputs. Dynamically biased amplifiers or integrators should not be used for accuracy.

It is also tempting to turn the power of high-Q oscillators or tank circuits on and off for low power. High-Q resonator is an energy storing element affected little by external source or load. That is, a Q-times higher energy is stored inside the tank. So it takes Q times longer cycles to fill up or drain the tank. The steady-state resonance cannot be initialized or jump-started instantly. Switching crystal oscillators is not desirable unless it is for long standby mode. Switching to RF tank circuits should be also avoided as it only generates transient distortion rather than shaping the sinusoidal resonance.

References

1. K. Kobayashi, N. Nogami, T. Shirotori, Y. Fujumoto, A current-mode latch sense amplifier and a static power saving input buffer for low-power architecture, in *Proceedings of VLSI Circuits Symposium*, June 1992, pp. 28–29
2. D. Schinkel, E. Mensink, E. Klumperink, E. van Tuijl, B. Nauta, A double-tail latch-type voltage sense amplifier with 18ps setup+hold time, *ISSCC Dig. Tech. Papers* (Feb 2007), pp. 314–315
3. B. Razavi, A strongARM latch. IEEE Solid State Circuits Mag **7**, 12–17 (2015)
4. W. Song, H. Choi, S. Kwak, B. Song, A 10-b, 20-MSamples/s low-power CMOS ADC. IEEE J. Solid State Circuits **30**, 514–521 (1995)
5. F. van der Goes, C. Ward, S. Astgimath, H. Yan, J. Riley, J. Mulder, S. Wang, K. Bult, A 1.5mW 68dB SNDR 80MS/s 2x interleaved SAR-assisted pipelined ADC in 28nm CMOS, *ISSCC Dig. Tech. Papers* (Feb 2014), pp. 200–201
6. A. Bugeja, B. Song, P. Lakers, S. Gillig, A 14b 100MS/s CMOS DAC designed for spectral performance. IEEE J. Solid State Circuits **34**, 1719–1732 (1999)
7. A. Bugeja, B. Song, A self-trimming 14b 100MS/s CMOS DAC. IEEE J. Solid State Circuits **35**, 1841–1852 (2000)
8. B. Razavi, A 2.4-GHz CMOS receiver for IEEE 802.11 wireless LAN's. IEEE J. Solid State Circuits **34**, 1382–1385 (1999)

Chapter 2
Continuous-Time Analog Circuits

Analog performance has been improved by careful design, new process, or feedback. Feedback is the only systematic way to enhance analog performance such as linearity, signal range, bandwidth, and impedance. Series (voltage) feedback increases linear voltage range while shunt (current) feedback increases linear current range. Also series feedback raises impedance while shunt feedback lowers impedance. Low impedance is for broad-banding while high impedance is for buffering. Feedback can be applied at any local or global levels, in continuous-time or discrete-time modes, and in broadband or DC servo applications.

2.1 Negative and Positive Feedbacks

There are two feedbacks. Negative feedback is to make stable systems such as power supplies, amplifiers, filters, $\Delta\Sigma$ modulators, PLL, and adaptive equalizers while positive feedback is to make unstable systems such as latches and oscillators. Only stability tells the difference between the negative and positive feedbacks. That is, negative-feedback amplifiers should be stable while positive-feedback oscillators should be unstable.

The feedback system is sketched conceptually in Fig. 2.1. The ports marked as i and o are the points the input is injected into and the output is taken from, and the path gains a and f represent the forward and feedback gains, respectively. Stability and dynamic performance are not affected by the locations of the input and output ports, but entirely by the loop gain of af. Depending on the polarity of the feedback loop gain, it makes either the negative or positive feedback.

Most closed-loop analog circuits except for latches and oscillators operate in stable negative feedback modes under the standard assumption of the small-signal, linear, time-invariant (SLT) operating condition. They benefit greatly from high linearity given by feedback. On the other hand, narrowband RF circuits such as low-noise amplifier (LNA) and mixer operate in open loop, and nonlinearity

B.-S. Song, *System-level Techniques for Analog Performance Enhancement*,
DOI 10.1007/978-3-319-27921-3_2

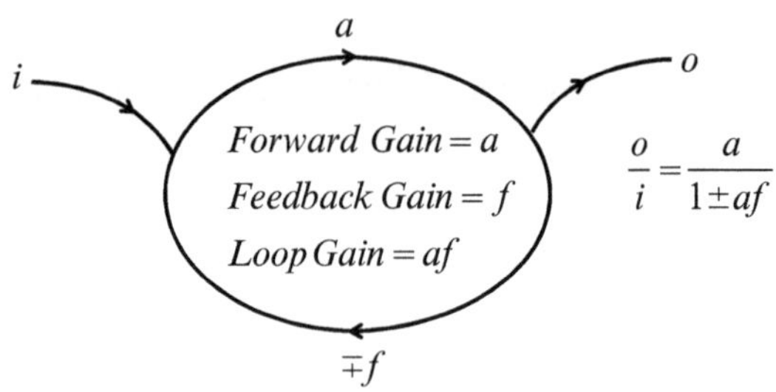

Fig. 2.1 Feedback system model

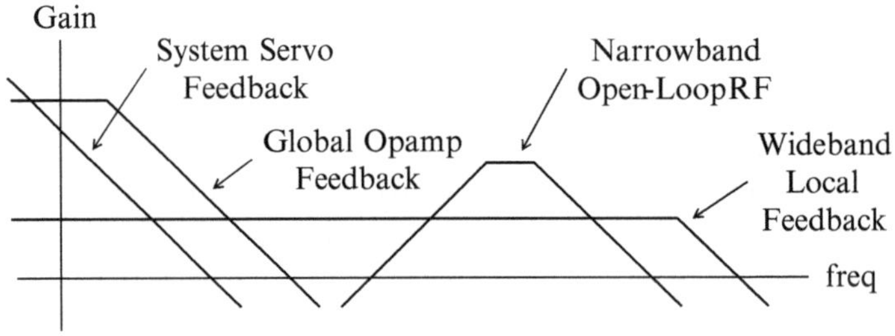

Fig. 2.2 Three kinds of analog circuits and feedbacks

becomes the most critical design constraint. RF design is basically how to achieve linearity while keeping noise low. However, in wideband baseband systems covering DC to GHz, local feedback can be applied to broadband amplifiers.

Figure 2.2 illustrates three operating frequency ranges of analog circuits. Feedback plays a critical role in obtaining the desired linearity except for narrowband RF circuits, which operate with inductor loads without feedback mainly to meet the low-power and low-noise requirements. Even at system levels, high-gain DC servo feedbacks can be applied so that performance can be enhanced by adapting to various circuit parameters.

Feedforward and feedback have quite different implications in circuits as shown in Fig. 2.3. The former is to feed the input forward and sum it at the output. The amplifier gain drops at high frequencies, but its output is held up by the signal fed forward instead. As a result, a zero is formed at the break point ω_z. However, since it is an open-loop implementation, there is no stability issue at all. On the other hand, the latter is to feed some of the amplifier output back and subtract it from the input. Since the subtracted difference is fed back into the amplifier, the input difference error is reduced by loop gain. That is, it makes a pole at the unity-loop gain frequency ω_k, where the loop gain becomes unity. If the extra phase delay of this amplified error approaches 180° at ω_k, the negative feedback becomes unstable positive feedback.

Feedforward is equivalent to the digital OR function as shown in Fig. 2.4. The analog OR function is to sum the responses of two parallel paths. In the wideband LNA, splitting noise into two parallel inverting and non-inverting paths and

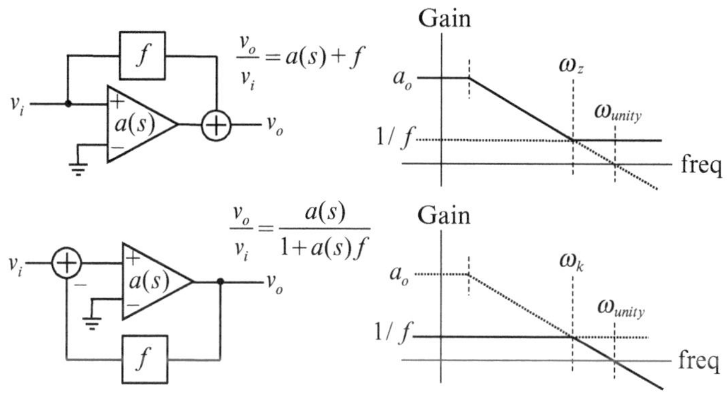

Fig. 2.3 Feedforward vs. feedback systems

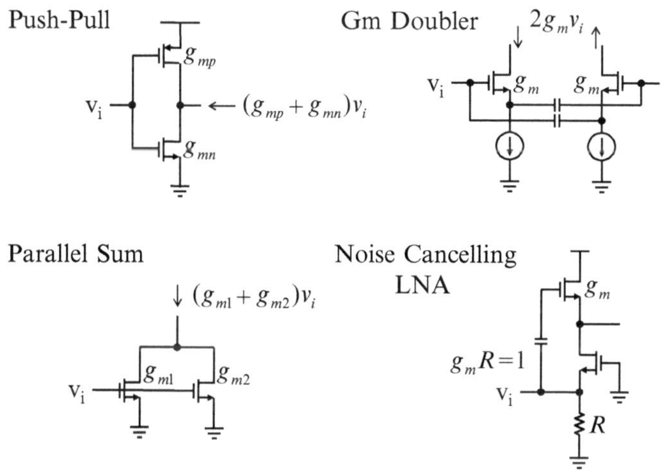

Fig. 2.4 Feedforward two-path examples

combining them also cancels the in-band noise—called noise cancellation. In addition, the sum of two parallel transistor currents effectively doubles the transistor g_m in the push–pull and parallel configurations. Although feedforward is useful to improve analog performance in some limited ways, it is only the negative feedback that can directly trade gain to enhance analog performance.

In the examples shown in Fig. 2.5, feedback loops encircles circuit elements not only in the voltage domain but also in the phase, frequency, and time domains. Depending on the error detector, the loop makes 3 unity-gain followers for phase, frequency, and time—called phase-locked loop (PLL), frequency-locked loop (FLL), and delay-locked loop (DLL), respectively. Among these, PLL is most widely used since the phase detector is the easiest among them to implement.

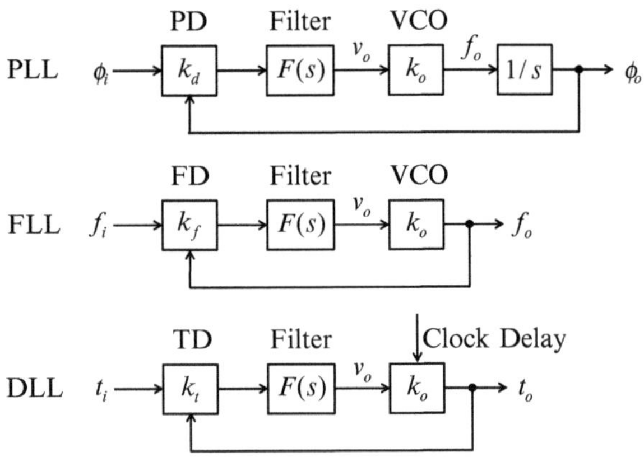

Fig. 2.5 Feedback examples of phase, frequency, and time followers

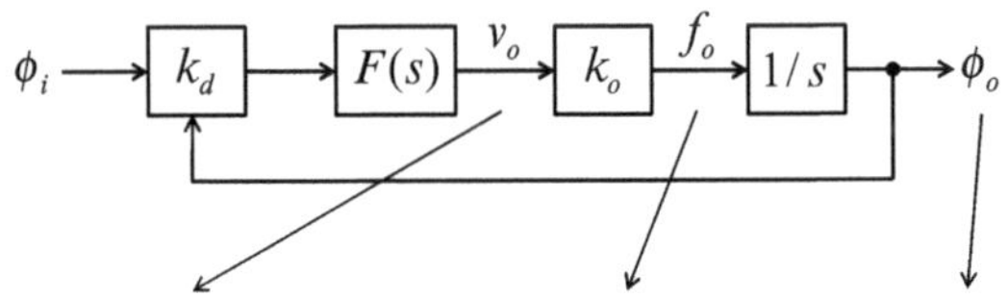

$$\frac{v_o}{\phi_i}=\frac{k_d F(s)}{1+k_d F(s)k_o/s},\quad \frac{f_o}{\phi_i}=\frac{k_d F(s)k_o}{1+k_d F(s)k_o/s},\quad \frac{\phi_o}{\phi_i}=\frac{k_d F(s)k_o/s}{1+k_d F(s)k_o/s}$$

$$\frac{f_o}{f_i}=\frac{\phi_o}{\phi_i}=\frac{k_d F(s)k_o}{s+k_d F(s)k_o} \Rightarrow \frac{f_o/k_o}{f_i}=\frac{v_o}{f_i}=\frac{k_d F(s)}{s+k_d F(s)k_o}$$

Fig. 2.6 Three forward gains with the same feedback loop gain

In the PLL, three outputs of voltage, frequency, and phase can be taken from the loop as shown in Fig. 2.6. The illustrated three transfer functions differ only in the forward gain, but they share the same loop gain. It is also shown that the voltage output is the demodulated FM output since the frequency is obtained by differentiating the phase, and the VCO converts voltage into frequency.

Most electronic systems use various local or global feedbacks. There are negative nonlinear feedback systems similar to PLL. An oversampling $\Delta\Sigma$ modulator suppresses quantization noise by loop gain. Manual trimming procedures carried out by engineers are also kinds of negative feedback schemes including human intelligence in the loop. However, the stability of all these loops is determined only by the loop gain and phase.

2.1.1 Phase Margin

In negative feedback, stability is determined by the phase margin at the frequency ω_k where the loop gain is unity or by the gain margin at the frequency ω_{180} where the excess loop phase is 180°. Both frequencies can be obtained from the following relations if the loop gain is set by $T(j\omega) = a(j\omega)f$.

$$|T(j\omega_k)| = 1 \quad \text{and} \quad \angle T(j\omega_{180}) = -180^\circ. \tag{2.1}$$

These two frequencies have special meanings: $\omega_k < \omega_{180}$ for stability, and $\omega_k > \omega_{180}$ for instability as graphically explained in the Bode plots shown in Fig. 2.7.

Both gain and phase margins (GM and PM) are defined as extra rooms for more loop gain and extra loop phase until the oscillation condition is reached. Feedback systems become unstable if the loop gain is greater than unity at the frequency where the total loop phase delay becomes the multiples of 2π such as 0°, 360°, and 720°. So for negative feedback systems to be stable, the PM should be positive, and the GM should be greater than 1. Similarly, for positive feedback systems, the PM should be negative, and the GM is smaller than 1. That is, referring only PM is sufficient to assure the stability of the feedback system. GM and PM are defined as follows.

$$\text{GM} = \frac{1}{|T(j\omega_{180})|} \quad \text{and} \quad \text{PM} = 180^\circ - |\angle T(j\omega_k)|. \tag{2.2}$$

Unless both GM and PM conditions are warranted, amplifiers get unstable and oscillate while oscillators get stable and amplify. That is, the boundary between the

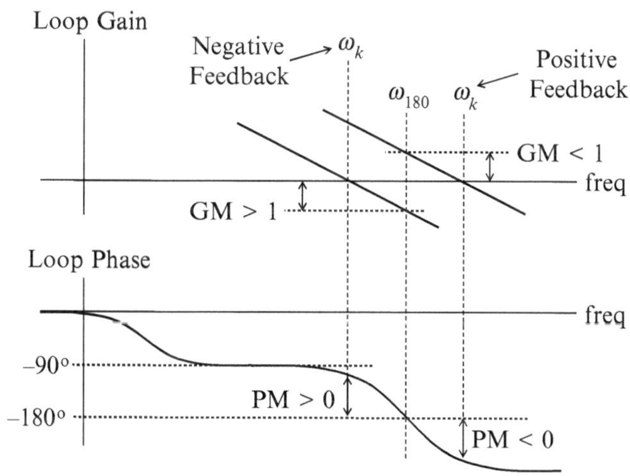

Fig. 2.7 Gain and phase margins for negative feedback

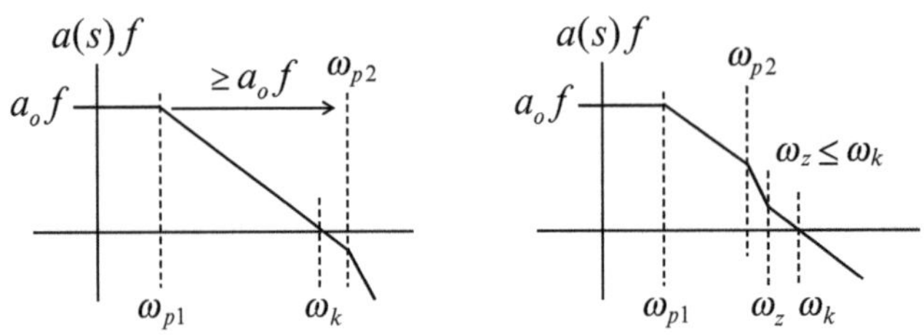

Fig. 2.8 Bode gain plots of two poles with and without a zero

negative and positive feedbacks is set by the loop gain at ω_{180}. If the loop gain is lower than unity, it is stable negative feedback. Otherwise, it becomes unstable positive feedback.

The stability condition can be stated in two ways as shown in the Bode gain plot of Fig. 2.8. The stability with PM greater than 45° is warranted if two poles are separated by more than the DC loop gain of $a_o f$. If two poles are not separated enough, a zero ω_z should be placed at lower frequencies than the unity loop-gain frequency ω_k. Then the PM becomes

$$\mathrm{PM} = 90^\circ + \tan^{-1}\frac{\omega_k}{\omega_z} - \tan^{-1}\frac{\omega_k}{\omega_{p2}}. \tag{2.3}$$

2.1.2 *Stability of Negative Feedback*

As noted, stability has been analyzed in many different ways. A few of them are: (1) Poles should be on the open left half plane. (2) The complex plot of the loop gain shouldn't encircle the (−1,0) point in negative feedback, and encircle the (1,0) point in positive feedback. (3) The zero-input response should die out as time goes by. (4) There should be GM and PM in the loop gain. Among them, the PM is most handy, and widely referred for stability. To relate the pole location and the PM, Root Locus which projects the trajectory of poles in the complex plane as a function of the feedback loop gain can be considered. The relation between the PM and the pole location for Chebyshev and Butterworth poles is illustrated for two-pole cases in Fig. 2.9.

If the unity loop-gain frequency ω_k is about the same as the second pole ω_{p2}, the PM is about 45°, and the complex conjugate poles are at 60° from the real axis in the Chebyshev case. On the other hand, if $\omega_k = \omega_{p2}/1.4$, then it makes the maximally flat Butterworth poles at 45° from the real axis. The closer to the imaginary axis the pole is, the higher the Q goes. High-Q complex conjugate pole pair causes peaking in the frequency response, and also makes overshoot and ringing in the transient response. The Root Locus shows the movement of poles as the loop gain is increased. The common two-pole feedback examples are shown in Fig. 2.10.

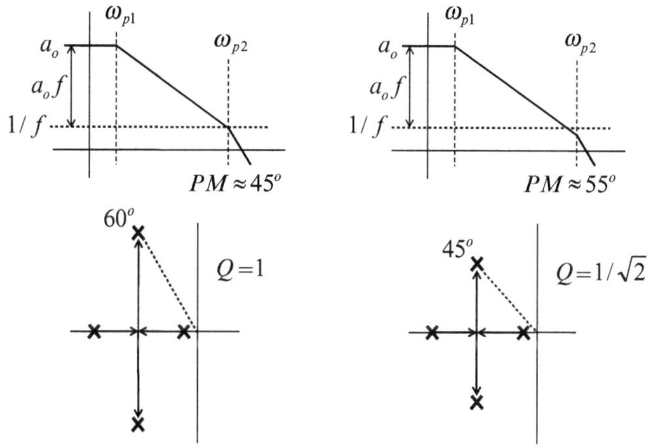

Fig. 2.9 Bode plots vs. poles in the complex plane

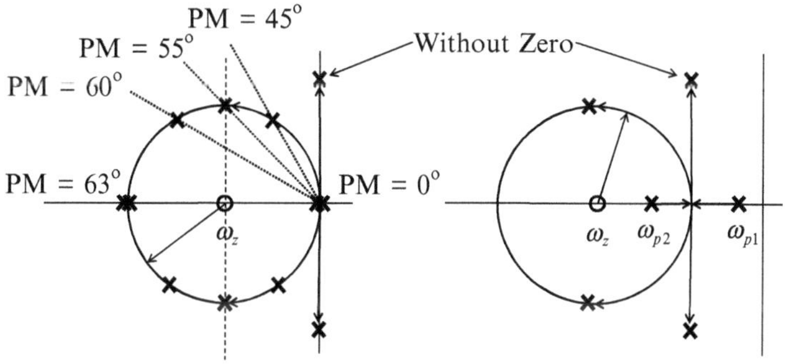

Fig. 2.10 Root Locus of two-pole feedback examples

Two open-loop poles enclosed in the feedback loop are two DC poles of integrators and two negative real poles such as in opamps, respectively. In both cases, two poles split into the vertical directions as the loop gain increases. A zero is needed to pull the poles into the open left half plane. Otherwise, as the loop gain further increases, the third pole easily pushes them into the right half plane and causes instability. The two DC poles split immediately while two negative real open-loop poles of ω_{p1} and ω_{p2} come closer before they split to be complex conjugate poles.

Let's consider two DC poles with a unity-gain frequency of ω_k. Due to the high DC gain, integrators are used as error amplifiers in most feedback systems such as PLL and $\Delta\Sigma$ modulator. If the unity-gain frequency is ω_k, the two-pole loop gain is $(\omega_k/s)^2$, and the closed loop transfer function has two imaginary axis poles at $+j\omega_k$ and $-j\omega_\kappa$. To move these poles into the open left half plane, a zero is placed at ω_z, where the loop gain is ω_k/ω_z. Then the open-loop gain becomes

$$\text{Loop Gain} = \frac{\omega_k}{s}\left(\frac{\omega_k}{s} + \frac{\omega_k}{\omega_z}\right). \tag{2.4}$$

Therefore, the closed-loop gain is given by

$$H(s) = \frac{\frac{\omega_k}{s}\left(\frac{\omega_k}{s} + \frac{\omega_k}{\omega_z}\right)}{1 + \frac{\omega_k}{s}\left(\frac{\omega_k}{s} + \frac{\omega_k}{\omega_z}\right).} \tag{2.5}$$

Here the gain factor is the ratio of ω_k/ω_z, which implies that the zero is placed at lower frequency than the unity-gain frequency by this ratio. By solving for the roots of the denominator polynomial, two poles can be found to be at

$$\left\{-\frac{\omega_k}{2\omega_z} \pm j\sqrt{1 - \left(\frac{\omega_k}{2\omega_z}\right)^2}\right\} \times \omega_k. \tag{2.6}$$

Two poles given by (2.6) are plotted in Fig. 2.10 as a function of ω_k/ω_z, and marked when its values are 0, 1, 1.414, 1.732, and 2. The PM can be also defined as follows.

$$\text{PM} = \tan^{-1}\left(\frac{\omega_k}{\omega_z}\right). \tag{2.7}$$

From (2.6), it can be shown that the Root Locus is also a circle centered at $-\omega_z$ with a radius of ω_z.

$$\sqrt{\left(-\frac{\omega_k^2}{2\omega_z} + \omega_z\right)^2 + \left\{1 - \left(\frac{\omega_k}{2\omega_z}\right)^2\right\} \times \omega_k^2} = \omega_z. \tag{2.8}$$

The Root Locus of two negative real poles with one zero as found in opamps is similar. The closed-loop gain with a DC gain of a_o and a loop gain of T_o is

$$H(s) = \frac{\frac{a_o\omega_{p1}\omega_{p2}}{\omega_z} \times \frac{(s+\omega_z)}{(s+\omega_{p1})(s+\omega_{p2})}}{1 + \frac{T_o\omega_{p1}\omega_{p2}}{\omega_z} \times \frac{(s+\omega_z)}{(s+\omega_{p1})(s+\omega_{p2})}}. \tag{2.9}$$

By solving for the roots of the denominator polynomial, two real poles are moved to

$$-\left(\frac{\omega_{p1} + \omega_{p1}}{2} + \frac{T_o\omega_{p1}\omega_{p2}}{2\omega_z}\right) \pm j\sqrt{(1 + T_o)\omega_{p1}\omega_{p2} - \left(\frac{\omega_{p1} + \omega_{p1}}{2} + \frac{T_o\omega_{p1}\omega_{p2}}{2\omega_z}\right)^2}. \tag{2.10}$$

Poles	X-Axis Angle	Phase Margin	Q	Band-Width
2 Real Poles	0°	63°	0.5	0.64
Bessel (Linear Phase)	30°	60°	0.58	0.8
Butterworth (Maximally Flat)	45°	55°	0.71	1
Chebyshev (1dB Ripple)	60°	45°	1	1.3

Fig. 2.11 Locations of two complex-conjugate poles

The Root Locus makes a circle around ω_z with a radius of the geometric mean of the distances to ω_{p1} and ω_{p2} from ω_z as follows.

$$\sqrt{\left(-\frac{\omega_{p1}+\omega_{p1}}{2}-\frac{T_o\omega_{p1}\omega_{p2}}{2\omega_z}+\omega_z\right)^2+(1+T_o)\omega_{p1}\omega_{p2}-\left(\frac{\omega_{p1}+\omega_{p1}}{2}+\frac{T_o\omega_{p1}\omega_{p2}}{2\omega_z}\right)^2}$$
$$=\sqrt{(\omega_z-\omega_{p1})(\omega_z-\omega_{p2})}. \tag{2.11}$$

Without ω_z, two poles split vertically at the middle frequency of $-(\omega_{p1}+\omega_{p2})/2$.

Figure 2.11 lists approximate the PM, Q, and -3 dB bandwidth for commonly used feedback amplifiers with different poles located with angles from the real axis: Two real poles, Bessel, Butterworth, and 1 dB-ripple Chebyshev. As two poles get closer to the imaginary axis, the Q gets higher, and both frequency and transient responses peak with larger overshoot. The maximally flat Butterworth response with poles at 45° gives about a PM of 55° while the higher-Q 1 dB ripple Chebyshev response with poles at 60° gives about a PM of 45°. Therefore, Bessel poles are required to design feedback amplifiers with a PM of over 60°.

Any negative feedback systems should be stabilized. Feedback amplifiers based on opamps are stabilized by separating two poles widely by more than the loop gain. If they are not separated widely or there are more non-dominant poles, inserting a zero is the way to get extra PM. However, for switched-capacitor applications that require accurate settling, inserting zero to cancel the phase delay of the second pole which is lower than the unity loop gain frequency should be avoided.

2.1.3 Instability of Positive Feedback

Positive feedback itself doesn't warrant instability though its purpose is to make systems like latches and oscillators unstable. Therefore, the oscillation condition for positive feedback systems to start with is

$$T(j\omega_{180}) = a(j\omega_{180})f > 1. \tag{2.12}$$

Unless the condition of (2.12) is met, even positive feedback stay stable, and oscillation will never grow. If met, the noise spectrum around ω_{180} will grow since the poles are on the right half plane. However, once the oscillation hits the voltage ceiling, the oscillation magnitude stops growing further, and the magnitude starts to be clipped and limited.

Therefore, the instability should be built into the design of a tuned oscillator as shown in Fig. 2.12. The limiting goes on and stops when the fundamental filtered by the tuned amplifier inside the loop meets the steady-state oscillation condition of

$$T(j\omega_{180}) = a(j\omega_{180})f = 1. \tag{2.13}$$

Due to the finite Q of the tuned amplifier, sidebands on both sides of ω_{180} grow too until limited, which makes the phase noise spectrum around the oscillation frequency.

Two examples of positive feedback are shown in Fig. 2.13. The bistable latch is unstable at DC if it meets the oscillation condition of (2.12). It settles back to one of the bistable states quickly with the g_m/C time constant. If the parasitic capacitance of the resistive load is tuned out with an inductor, it makes an unstable oscillator at a resonant frequency if the same condition of (2.12) is met. Once variable capacitor is added, the standard VCO biased with a tail current is obtained as shown on the right side.

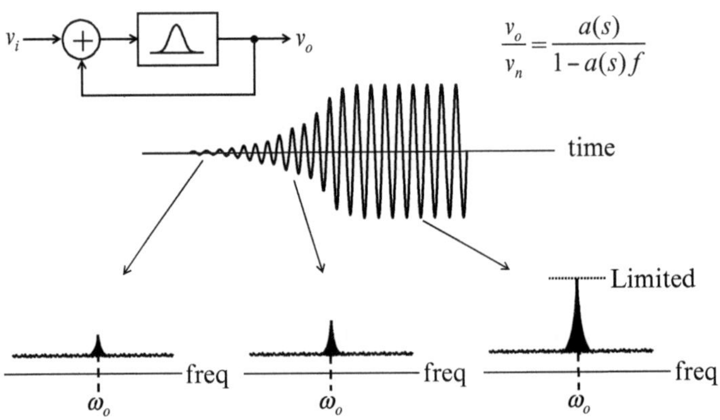

Fig. 2.12 Positive feedback example

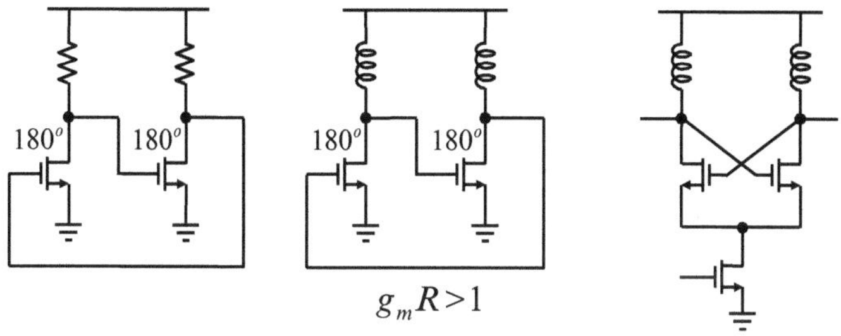

Fig. 2.13 Latch and VCO

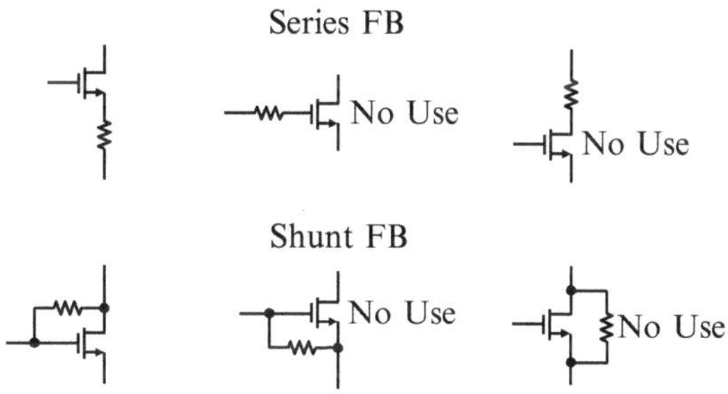

Fig. 2.14 Series and shunt feedback examples using a resistor

2.2 Local Series and Shunt Feedbacks

Feedback comes in two different forms: Series and shunt. The former is the voltage feedback while the latter is the current feedback. Therefore, the former widens linear voltage range and raises impedance while the latter widens linear current range and lowers impedance. They are dual in concept. Simple feedback circuits can be made using one transistor and one passive component such as resistor, capacitor, and inductor.

All six combinations possible with a resistor are shown in Fig. 2.14. Among them, only source degeneration and shunt feedback are useful for local feedbacks. Other four configurations are not necessary. Similarly, only three useful feedbacks are possible with an inductor and a capacitor as shown in Fig. 2.15.

In the inductor source degeneration, the driving-point input resistance becomes real when the inductor series-resonates with the gate-source capacitance, which is used as a matching load to antenna in the LNA design. Examples of the shunt feedback using an inductor can be found in the Colpitts and Pierce oscillator designs, and the capacitive shunt feedback is an integrator often used to make a Miller capacitance.

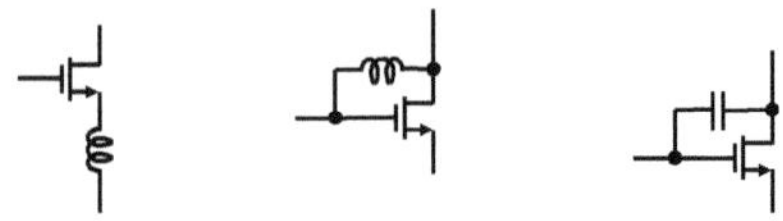

Fig. 2.15 Other useful shunt and series feedback examples

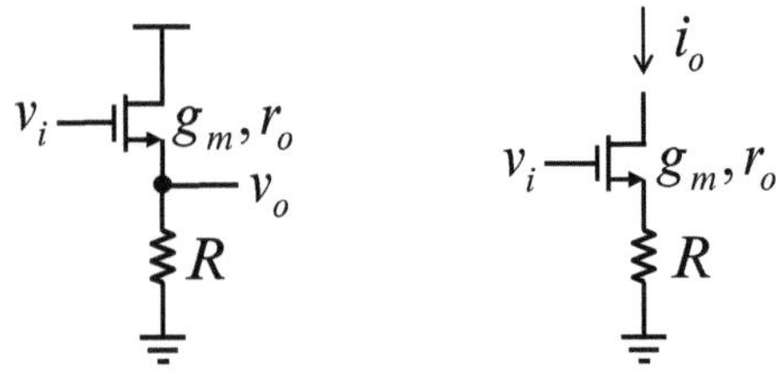

Fig. 2.16 Two examples of series feedback

2.2.1 *Series Feedback*

The only useful series feedback using a resistor is the source degeneration. In BJT circuits, transistors are rarely used without emitter degeneration since the input resistance looking into the base is low limited by the current gain. On the other hand, the source degeneration for MOS transistors is not necessary, but it still makes a useful circuit configuration for the following two cases: Source follower and source degeneration as shown in Fig. 2.16.

Assuming two small-signal parameters are g_m and r_o, the low-frequency small-signal closed-loop gain v_o/v_i, the forward gain a_o, the feedback gain f, and the loop gain a_of can be derived for the source follower shown on the left side as follows.

$$\begin{aligned} \frac{v_o}{v_i} &= \frac{g_m(r_o \| R)}{1 + (g_m - g_{mb})(r_o \| R)} \approx \frac{g_m}{g_m - g_{mb}}, \\ a_o &= g_m(r_o \| R), \quad f = \frac{g_m - g_{mb}}{g_m}, \\ a_of &= (g_m - g_{mb})(r_o \| R), \end{aligned} \tag{2.14}$$

respectively. Unlike the BJT emitter follower, the small-signal gain of the MOS source follower is lower than unity due to the body g_{mb}, which is about 10–20 % of g_m.

The trans-conductance of the MOS transistor with the source degeneration works similarly, but the output is current drawn from the high-impedance drain side. The total trans-conductance i_o/v_i, the forward gain a_o, the feedback gain f, and the loop gain a_of can be also derived as follows.

$$\frac{i_o}{v_i} = \frac{g_m}{1 + (g_m - g_{mb})(r_o \| R)},$$
$$a_o = g_m, \quad f = \frac{g_m - g_{mb}}{g_m}(r_o \| R), \tag{2.15}$$
$$a_o f = (g_m - g_{mb})(r_o \| R),$$

respectively. Note that g_m decreases by the amount of the loop gain in series feedbacks, but its linearity improves by the same amount instead. Both cases are the same series feedback with the same loop gain. The loop gain is irrelevant of input and output ports.

In most bulk N-well processes, all NMOS bodies are tied to one substrate. In analog circuits which are not switched, the body effect of the PMOS transistor can be alleviated if the substrate is tied to its source. However, in digital circuits, even PMOS substrates are tied to the high supply voltage. If the source is floating with the body tied to the supply, the body effect raises the effective g_m to be $(g_m - g_{mb}) = (1.1\text{–}1.2)g_m$. Furthermore, since the output resistance r_o of the MOS transistor is not as high as that of bipolar transistors, the gain of the NMOS source follower is even lowered to the 0.8–0.9 level, and rarely approaches unity.

2.2.2 Source Follower

The source follower is a unity-gain voltage buffer. The basic function of the buffer is to transform impedance from high to low. That is, it is a light load to the input, but its low output impedance is to drive a heavy load. Its input impedance is high and mostly capacitive. Therefore, at low frequencies, the high-impedance input is open, but as frequencies go higher, the input capacitance looking into the gate is given by the Miller effect as follows.

$$C_i = (1 - A)C_{gs} + C_{gd} \approx C_{gd}, \tag{2.16}$$

where A is the source follower gain. That is, if $A = 0.9$, only about 10 % of C_{gs} loads the input. On the other hand, its output driving-point resistance is low.

$$R_o = \frac{1}{g_m - g_{mb}} \| r_o \| R \approx \frac{1}{g_m - g_{mb}}. \tag{2.17}$$

This characteristic of high input and low output impedances stays valid up to the almost device unity-gain frequency.

At high frequencies, the feedforward zero of the MOS transistor is always at g_m/C_{gs}. Note that the gate-source feedforward zero makes a negative real zero while the gate-drain feedforward zero creates a positive real zero due to the polarity inversion of the signal path. After the zero frequency, signal just bypasses the

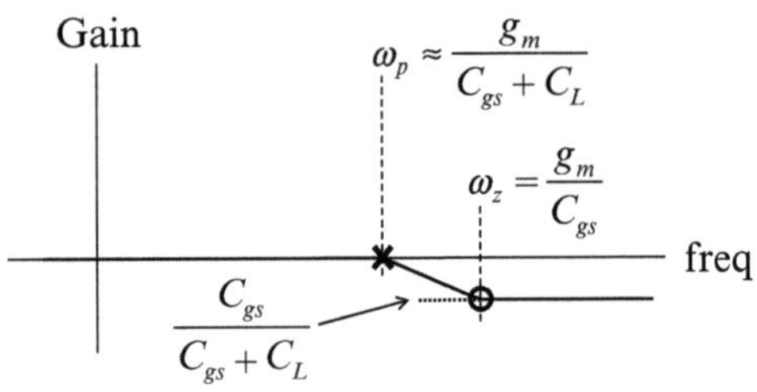

Fig. 2.17 Frequency response of the source follower

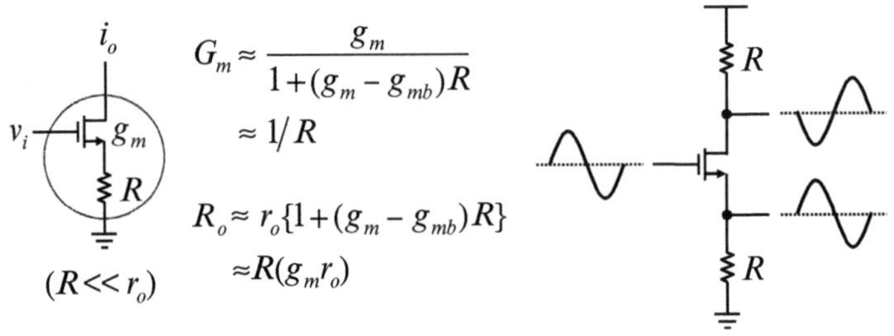

Fig. 2.18 Phase splitter using source degeneration

transistor, goes through C_{gs}, and drives the loading capacitance C_L directly, which implies that the high-frequency attenuation converges to the capacitor divider ratio of $C_{gs}/(C_{gs}+C_L)$. Therefore, the pole and the zero are separated by this gain factor, and the pole frequency is lower as sketched in Fig. 2.17.

Therefore, the frequency response can be derived as

$$\frac{v_o}{v_i}(s) = \frac{g_m R\left(1 + sC_{gs}/g_m\right)}{1 + (g_m - g_{mb})R + sR\left(C_{gs} + C_L\right)}. \tag{2.18}$$

If r_o is ignored and s is set to 0, it is the same equation as (2.14). The feedforward zero g_m/C_{gs} of the source follower affects its high-frequency performance as it provides a leaky forward path for signal. The feedforward signal leak is very troublesome when using it as a unity-gain buffer such as in Sallen-Key type low-pass filters that require unity-gain buffers. The output noise over wide bandwidth is also a problem when it is aliased into the signal band when the filter output is sampled.

Since the same current flows through the MOS transistor, the small-signal source and drain voltages are 180° out of phase from each other. Therefore, an RF phase splitter can be made as shown in Fig. 2.18. The source-degenerated MOS transistor encircled by the dotted line exhibits the trans-conductance of approximately $1/R$ while its driving-point output resistance is improved by the common gain factor of $g_m r_o$ as approximated.

2.2.3 *Inductor Source Degeneration*

The g_m of MOS transistor is usually low, but has a wider linear range than that of BJT. If degenerated with R for series feedback, the trans-conductance further decreases by the amount of the loop gain $\{1+(g_m-g_{mb})R\}$. Therefore, the input linear range of the MOS transistor is widened, and its input capacitance also gets smaller by the same factor. That is, the transistor characteristic approaches that of an ideal transistor, which has high input and output impedances plus linearized trans-conductance.

Except in broadband networking systems such as Giga-bit Ethernet and fiber, most RF circuits operate in open-loop conditions without feedback. Due to low g_m, the source degeneration is not common in low-frequency circuits such as in opamps, but it is often used in RF open-loop circuits if high linearity is required such as in the LNA and mixer. In RF circuits, inductors tune out parasitic capacitances.

Figure 2.19 shows the standard LNA with the input impedance matched to the source impedance. The source degeneration given by the inductor L makes an inductive trans-conductance device.

$$\frac{i_o}{v_i}(s)=\frac{g_m}{1+(g_m-g_{mb})\times sL+s^2LC_{gs}}=\frac{g_m}{(g_m-g_{mb})}\times\frac{1}{sL}, \tag{2.19}$$

at the resonant frequency where $s^2=-1/LC_{gs}$. It is unusual to use it as a source follower with inductor degeneration as a buffer, but the source follower gain can be obtained as

$$\frac{v_o}{v_i}(s)=\frac{g_m\times sL(1+sC_{gs}/g_m)}{1+(g_m-g_{mb})\times sL+s^2LC_{gs}}, \tag{2.20}$$

which is the same as (2.18) with the same feedforward zero at g_m/C_{gs}. When the series L and C_{gs} resonate, a real resistance remains. Let's get the voltage drop (v_i-v_o) across the gate and source. From (2.20), we obtain

$$\frac{(v_i-v_o)}{v_i}(s)=\frac{1}{1+(g_m-g_{mb})\times sL+s^2LC_{gs}}. \tag{2.21}$$

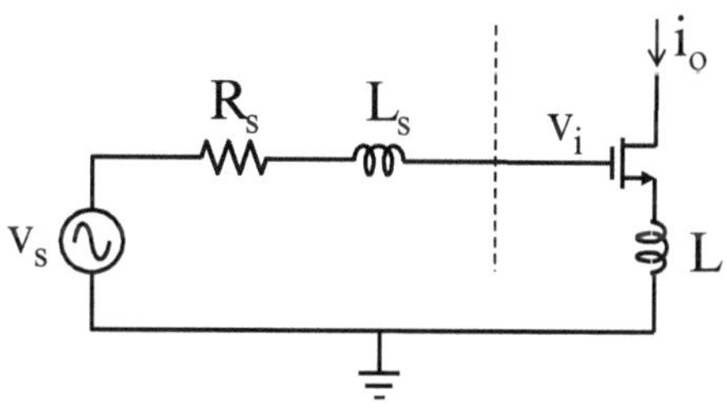

Fig. 2.19 Impedance-matched LNA with inductor source degeneration

Then from (2.21), the input impedance looking into the gate is given by

$$\frac{v_i}{i_i}(s) = \frac{v_i}{sC_{gs}(v_i - v_o)} = \frac{1 + (g_m - g_{mb}) \times sL + s^2 LC_{gs}}{sC_{gs}} = \frac{(g_m - g_{mb})L}{C_{gs}}, \quad (2.22)$$

again at the resonant frequency. This is the real resistance which can terminate the input with impedance matched to R_s. For LNA, noise figure (NF) improves with larger g_m. Therefore, the inductor L can be minimized, and an extra inductance L_s is inserted to make a resonance while keeping the impedance matched.

$$\omega_o = \frac{1}{\sqrt{(L + L_s)C_{gs}}}, \quad R_s = \frac{(g_m - g_{mb})L}{C_{gs}}. \quad (2.23)$$

Therefore, the LNA design is straightforward. $(L + L_s)$ tunes out C_{gs}, and the residual resistance $R_L + R_{Ls} + R_g + (g_m - g_{mb})L/C_{gs}$ can be matched to R_s, which is typically 50 Ω, where physical inductor and transistor gate resistances are included. Since the dominant $(g_m - g_{mb})L/C_{gs}$ doesn't contribute to noise directly, and NF can go below 3 dB. If the MOS noise is referred to the input, the NF can be approximated as follows.

$$\mathrm{NF} = 1 + \frac{2}{3} \times \frac{(\omega_o C_{gs})^2 R_s}{g_m} = 1 + \frac{2}{3} \times \left(\frac{\omega_o}{\omega_T}\right)^2 \times g_m R_s, \quad (2.24)$$

where ω_T is the device unity-gain frequency defined as g_m/C_{gs}. Large g_m/C_{gs} obtained by device scaling lowers the NF.

The gate resistance R_g can be made small by careful layout, and the series inductor resistances R_L and R_{Ls} are small. Therefore, after C_{gs} is tuned out, and $(g_m - g_{mb})L/C_{gs}$ is matched to R_s, the effective total g_m becomes

$$G_m = \frac{i_o}{v_s} = \frac{i_i \times \frac{g_m}{sC_{gs}}}{2v_i} = \frac{i_i \times \frac{g_m}{sC_{gs}}}{2v_i} = \frac{\omega_T}{\omega_o} \times \frac{1}{2R_s}, \quad (2.25)$$

which is independent of the device g_m. That is, technology scaling with smaller input capacitance will increase G_m. Therefore, the LNA design is almost set once technology is given: (1) Set the trans-conductance g_m for noise. (2) Set the overdrive voltage $(V_{gs} - V_{th})$ for the intercept point and linearity. Then, the device size and bias current are set. (3) Estimate the gate resistance. (4) Select L for matching. (5) Select L_s for input resonance. (6) Check the total resistance by estimating series resistances of L and L_s, and iterate the procedure if needed. More power is consumed with non-ideal factors such as input pad parasitic and Miller capacitances, time-variant channel charge, and hot carrier effects. Usually better NF is observed with the input resistance set lower than R_s.

2.2.4 Resistance Reflection in Series Feedback

Series voltage feedback raises impedance by the amount of the loop gain. The series resistance R can be inserted in the source and drain branches as shown in Fig. 2.20. If the body effect given by g_{mb} is ignored for simplicity, the resistance looking into the drain and source sides increases or decreases by the same amount of $g_m r_o$, which is the maximum gain obtainable from one transistor amplifier. This value of 20–40 dB varies depending on the process, channel length, and bias condition. The resistances looking into the drain and the source are r_o and $1/g_m$, respectively, without source degeneration, but they change to $R(g_m r_o)$ and $R/(g_m r_o)$, respectively. Therefore, from the drain side, the source-side resistance looks larger, but from the source side, the drain-side resistance looks smaller by the same factor [1].

The driving-point resistances of dual and triple cascode circuits are generalized in Fig. 2.21 assuming again that all devices have the same g_m and r_o. The resistance value R can be an output resistance r_o of another transistor. The highest resistance level possible in MOS circuits is limited by leakage. The output resistance offered by the triple cascode is getting closer to the highest impedance node limited by the junction leakage.

The cascode examples to get higher gain in opamps are shown in Fig. 2.22. Cascoding raises the output resistance by $g_m r_o$, thereby enhancing the gain by the same factor. The high output resistance when seen from the cascoded node is reduced by the same factor. That is, the driving-point resistance of the cascoded

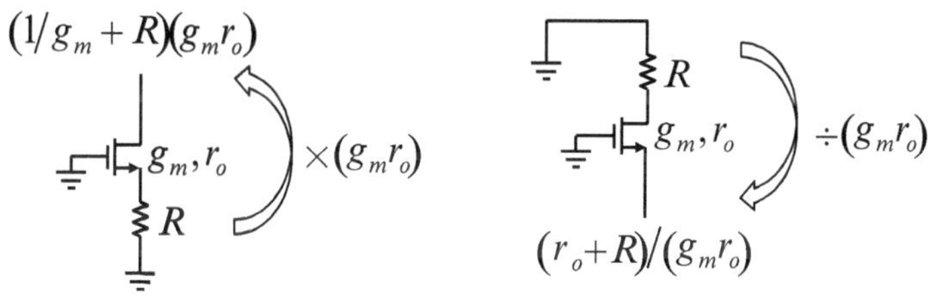

Fig. 2.20 Two useful resistance reflection rules

Fig. 2.21 Resistance reflection rules generalized

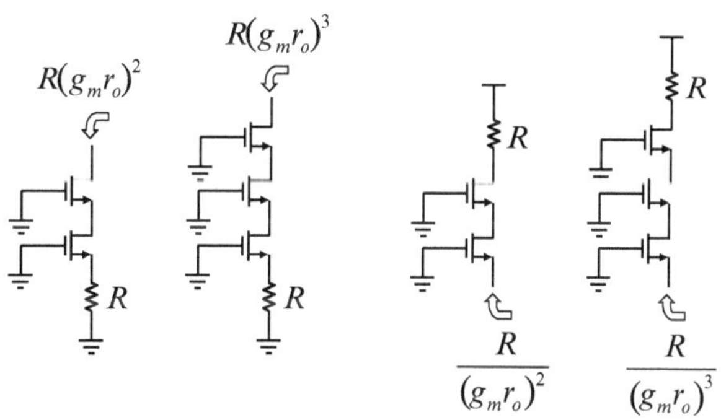

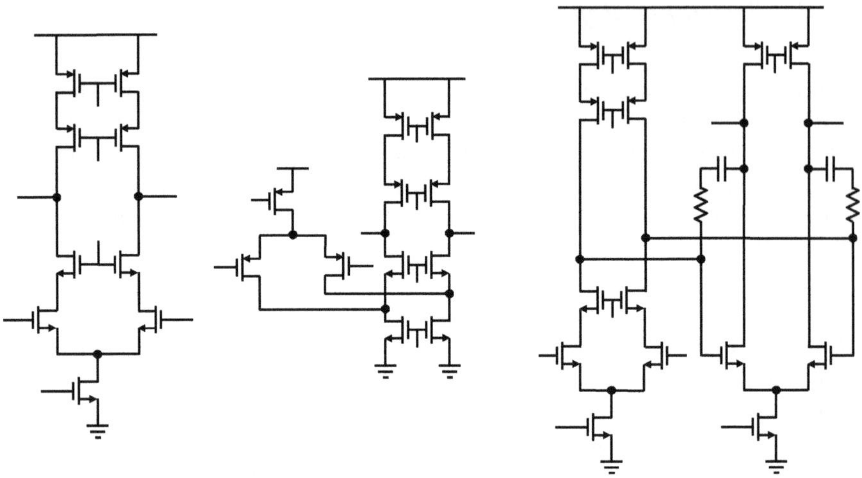

Fig. 2.22 Single-stage cascode, folded-cascode, and two-stage opamps

node approaches $1/g_m$ at high frequencies since the output node is loaded by the large capacitance to make a dominant pole. Therefore, cascoding only adds a very high frequency non-dominant pole at g_m/C_p with a parasitic capacitance C_p.

Standard opamp designs have been well established. Although the impedance level is raised for high gain by cascoding, it is difficult to cascode with low supply voltages since it requires additional DC voltage drop across the cascode device. At low-voltage uses, two-stage opamps have been preferred to single-stage opamps. In switched-capacitor applications, the input stage can be cascoded for high gain while the second stage gives high swing. In high-swing buffer applications that require high input-common mode voltages, either rail-to-rail or class AB input stages are used. However, in scaled technologies, supply voltages are still tight even for double cascoding.

To get higher gain without using multiple cascoding, a gain boosting technique based on shunt feedback can be used as sketched in Fig. 2.23. One problem that comes with the local shunt feedback is that the unity loop-gain frequency of the local feedback loop becomes a zero in the main gain path. The doublet effect on settling can be alleviated by moving the zero to higher frequencies than the unity-gain frequency of the main loop.

2.2.5 *Shunt Feedback*

When compared to the series voltage feedback, some of the output current is fed back to the input in the current shunt feedback, thereby reducing both the input and output driving-point impedances. The only useful shunt feedback with a resistor is the trans-resistance configuration with a resistor connected between the gate and the drain. It is compared to the source-degeneration series feedback in Fig. 2.24.

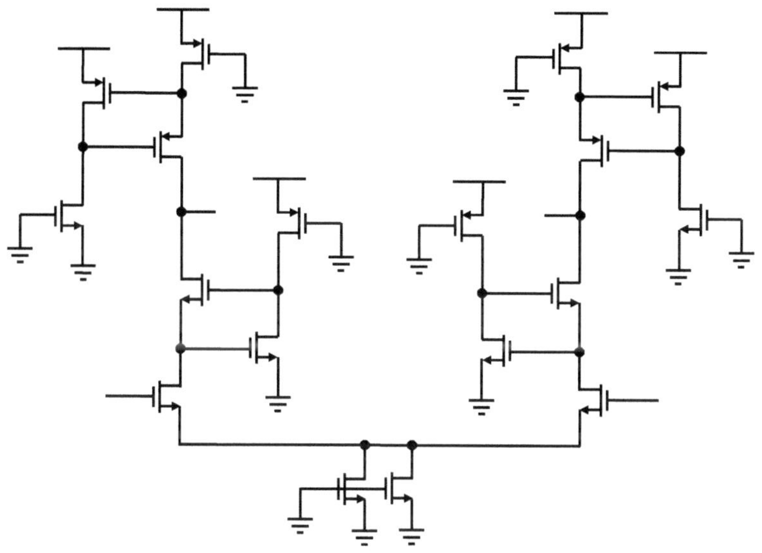

Fig. 2.23 Gain boosting example by shunt feedback

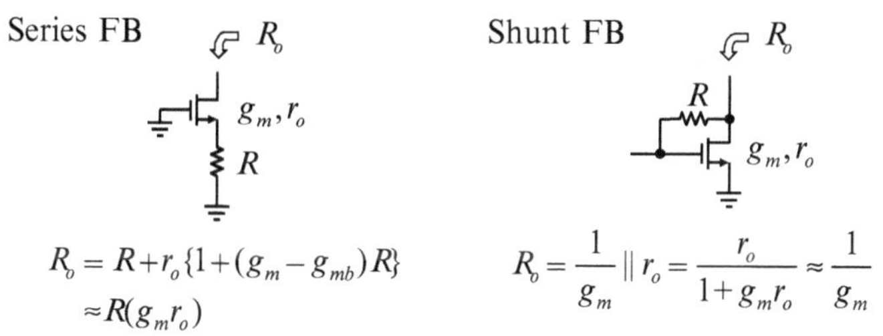

Fig. 2.24 Comparison between series and shunt feedbacks

The output resistance of the shunt feedback is the parallel combination of the diode resistance and the transistor output resistance since the gate-side resistance of the MOS transistor is infinite. That is, the drain and the gate are shorted at low frequencies.

Figure 2.25 summarizes the resistance reflection rules for shunt feedbacks. The driving-point resistances looking into the output and input ports are the same diode resistance $1/g_m$ plus shunt resistance R divided by the loop gains of $g_m R_S$ and $g_m R_L$, respectively. This low resistance offered by the shunt feedback helps to broadband amplifiers. That is why the standard trans-resistance amplifier has been used for benchmarking new high-speed technologies. It has also been used to amplify low light-sensitive currents from photo diodes since it provides low impedance load to photo detector current.

This symmetry of the input and output resistance reflection rule gives a hint that the driving-point resistances looking into the input and output ports can be matched.

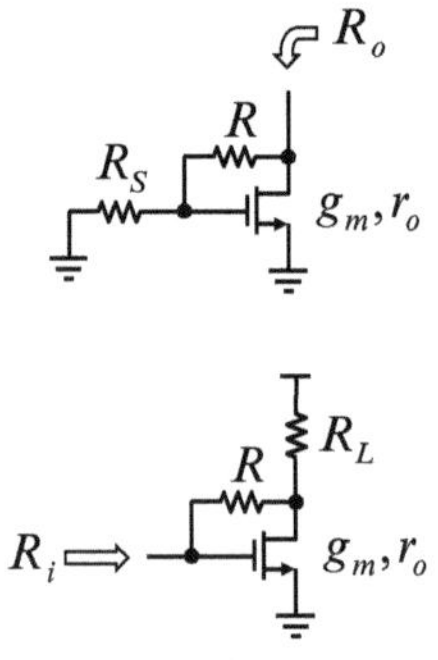

$$R_o = (R+R_S)\|\frac{R+R_S}{g_m R_S}\| r_o \approx \frac{R+R_S}{1+g_m R_S} \approx \frac{1}{g_m} + \frac{R}{1+g_m R_S}$$

$$R_i = \frac{R+R_L\|r_o}{1+g_m(R_L\|r_o)} \approx \frac{R+R_L}{1+g_m R_L} \approx \frac{1}{g_m} + \frac{R}{1+g_m R_L}$$

Fig. 2.25 Resistance reflection rules for shunt feedbacks

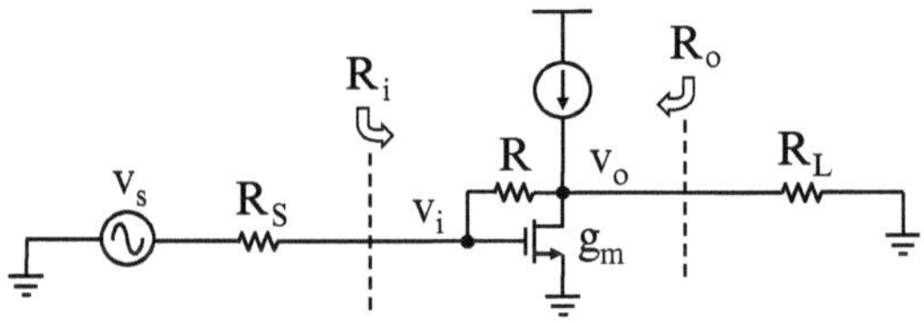

Fig. 2.26 Impedance-matched wideband amplifier

That is, wideband amplifiers with both matched input and output resistances can be implemented using a trans-resistance amplifier as shown in Fig. 2.26.

For input and output driving-point resistances to be matched, the following condition should be met.

$$R_S = R_i = \frac{R+R_L}{1+g_m R_L} = \frac{R+R_S}{1+g_m R_S} = R_o = R_L, \tag{2.26}$$

which gives the simple relation of

$$R = g_m R_S R_L. \tag{2.27}$$

Due to this symmetric matching characteristic, any local shunt-feedback stages can be cascaded for higher gain. Shunt feedback also improves linearity while lowering the resistance level for broadbanding. The small-signal voltage gain of the shunt-feedback stage can be derived as follows.

$$\frac{v_o}{v_i} = -\frac{g_m R R_L - R_L}{R+R_L} \approx -\frac{g_m R R_L}{R+R_L} = -g_m(R\|R_L). \tag{2.28}$$

Similarly, the gain including R_S is obtained.

$$\frac{v_o}{v_s} = -\frac{g_m R R_L - R_L}{R+R_S+R_L+g_m R_S R_L} \approx -\frac{g_m R R_L}{R+g_m R_S R_L}. \tag{2.29}$$

Alternatively, (2.29) can be approximated in two steps of attenuation and amplification as follows.

$$\frac{v_o}{v_s} = \frac{v_i}{v_s} \times \frac{v_o}{v_i} = -\frac{\dfrac{R+R_L}{1+g_mR_L}}{R_S + \dfrac{R+R_L}{1+g_mR_L}} \times \frac{g_mRR_L - R_L}{R+R_L} \approx -\frac{g_mRR_L}{R+g_mR_SR_L}. \tag{2.30}$$

At high frequencies, the input and output driving-point resistances can be matched to the standard 50 Ω, which facilitates its use as an amplifier with both the input and output ports loaded by transmission lines as shown in Fig. 2.27.

If $R_i = R_o = 50\ \Omega$, we obtain the following from (2.27) and (2.28).

$$\begin{aligned} R &= g_mR_SR_L = 2500g_m, \\ \frac{v_o}{v_s} &\approx -\frac{g_mRR_L}{R+g_mR_SR_L} = -\frac{g_mR_L}{2} = -25g_m = -\frac{R}{100}. \end{aligned} \tag{2.31}$$

Fog higher gain, both g_m and R should be set higher as follows.

$$\begin{aligned} g_m &= 1/10\ \Omega, \quad R = 250\ \Omega, \quad \text{Gain} \approx -2.5. \\ g_m &= 1/5\ \Omega, \quad R = 500\ \Omega, \quad \text{Gain} \approx -5. \\ g_m &= 1/2.5\ \Omega, \quad R = 1\ \text{k}\Omega, \quad \text{Gain} \approx -10. \end{aligned} \tag{2.32}$$

Impedance matched amplifiers are mostly to drive transmission lines or antenna ports such as in monolithic microwave integrated circuits or RF transceivers. For integrated on-chip uses, it is not necessary to match impedance for amplifier input or output ports since there are no transmission lines. Therefore, either current or voltage source is required to drive high or low impedance load, respectively, as shown in Fig. 2.28.

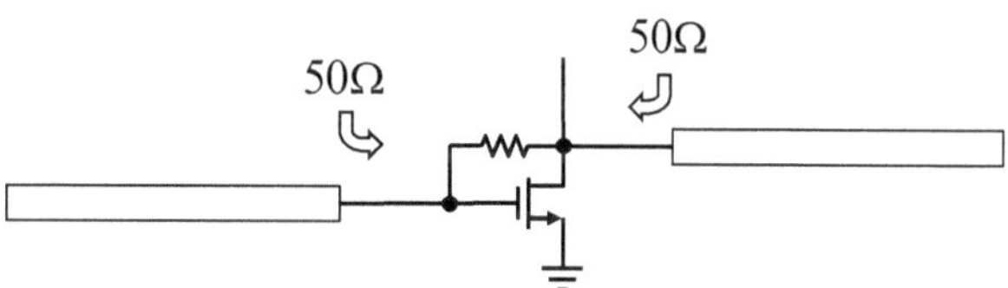

Fig. 2.27 Amplifier with matched input and output ports

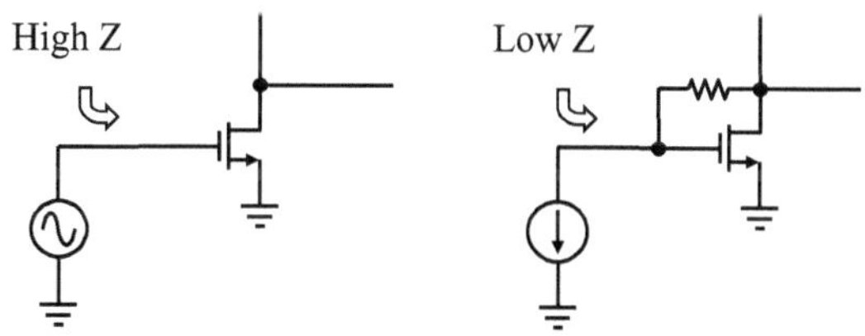

Fig. 2.28 Sources for high and impedance nodes

The stability of local series or shunt feedback is not an issue in general since only one dominant pole is involved. However, in shunt feedback, as the shunt resistance value increases, input and output nodes are separated, and make two poles in the loop. It sets the upper bound to the maximum bandwidth achievable using shunt-feedback amplifiers such as trans-resistance amplifiers. In particular, the capacitance at the input node is very critical since it lowers the pole frequency in the feedback path. The impedance at the input node is affected differently by the shunt feedback. The resistive shunt feedback lowers the input resistance while the capacitive shunt feedback increases the input capacitance due to the Miller effect as shown in the two cases of Fig. 2.29.

Note that unlike the voltage-driven opamp case, the effective gain of the shunt-feedback transistor amplifier decreases by the loading of the shunt feedback resistance. The shunt resistance from the input side looks smaller by the shunt-stage gain while the shunt capacitor looks larger by the same amount. The former helps to broadband the frequency response while the latter does the opposite as shown in Fig. 2.30.

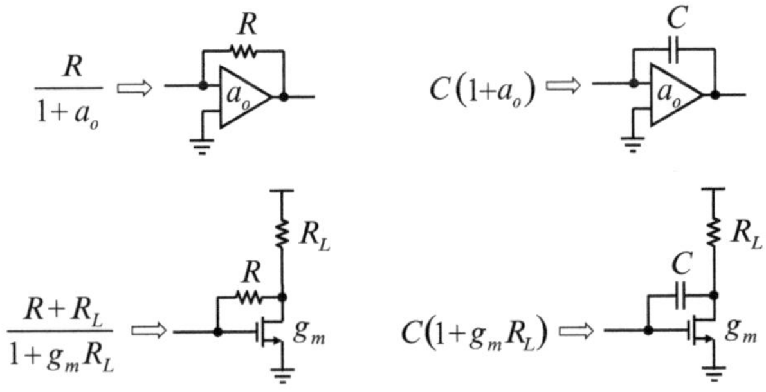

Fig. 2.29 Shunt feedback vs. Miller effect

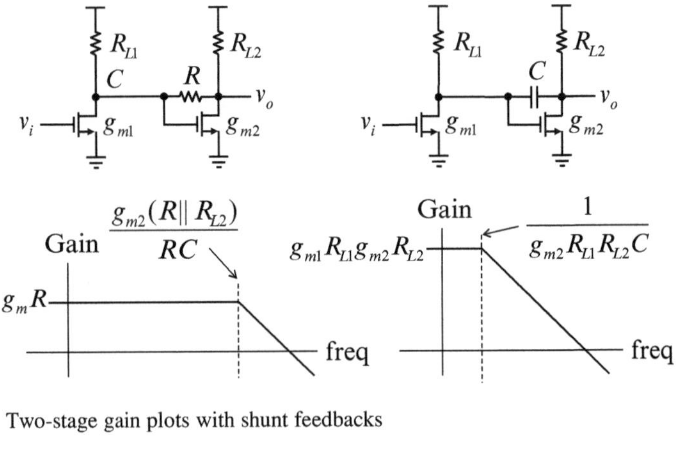

Fig. 2.30 Two-stage gain plots with shunt feedbacks

Broad-banding by shunt feedback is possible as the resistance level drops. However, its upper limit is set by the non-dominant second pole at the output node if this pole is pushed out too high. On the other hand, the Miller effect is used to frequency-compensate two-stage opamps by moving the dominant pole lower and separating two poles widely, which is called narrow-banding.

2.3 Trans-Resistance Amplifier

If the shunt-feedback amplifier is driven by the current source, it makes another useful local feedback circuit like the local series-feedback source degeneration. The input and output resistances are lowered by shunt feedback while they are raised by the series feedback. Its input resistance, voltage gain, and trans-resistance can be obtained as summarized in Fig. 2.31.

That is, the input current makes the voltage drop at the input of the trans-resistance stage, which is amplified by the voltage gain stage. Since the gain is defined as the ratio of the output voltage to the input current which has the resistance unit, it has been called trans-resistance amplifier.

Figure 2.32 illustrates the logic behind the preamplifier issue in optical receivers. Photo diodes generates low-level currents of a nA ~ μA order depending on the intensity of the incident light. To convert it into voltage, a resistor is needed to develop a voltage across it. If there is a parasitic C_D of the detector, the impedance drops after the pole frequency at $1/RC_D$. If an amplifier drives the resistance in the shunt-feedback form, the bandwidth can be widened by the loop gain $(1 + a_o)$. The parasitic capacitance C_D at the detector input node is the most important parameter to consider in the trans-resistance amplifier design. For higher output current, the size of photo detector should be made large. Then large diode gives large parasitic capacitance C_D.

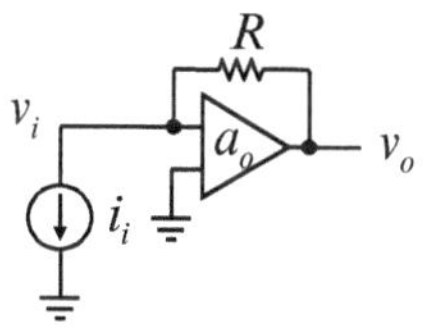

$$\frac{v_i}{i_i} = \frac{R}{1+a_o}, \quad \frac{v_o}{v_i} = a_o$$

$$\frac{v_o}{i_i} = \frac{v_i}{i_i} \times \frac{v_o}{v_i} = \frac{a_o R}{1+a_o} \approx R - \frac{R}{a_o}$$

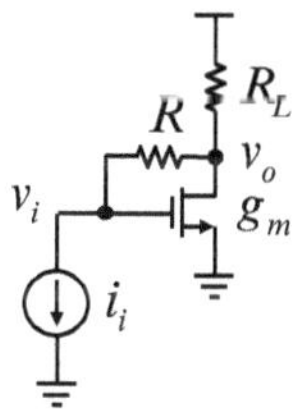

$$\frac{v_i}{i_i} = \frac{R \,|\, R_L}{1+g_m R_L}, \quad \frac{v_o}{v_i} = \frac{g_m R R_L \;\; R_L}{R+R_L}$$

$$\frac{v_o}{i_i} = \frac{v_i}{i_i} \times \frac{v_o}{v_i} = \frac{g_m R R_L - R_L}{1+g_m R_L} \approx R - \frac{1}{g_m}$$

Fig. 2.31 Trans-resistance amplifier

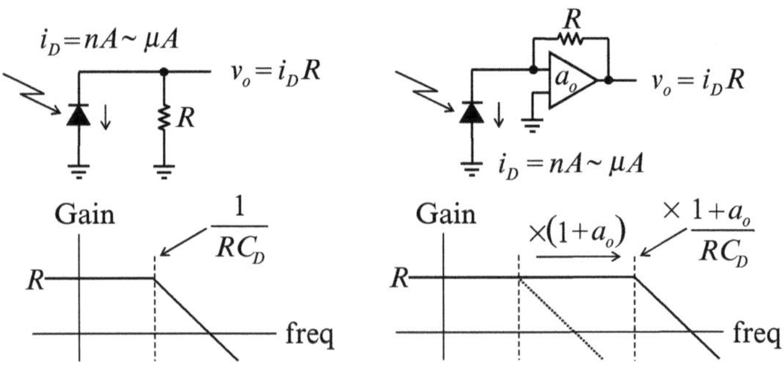

Fig. 2.32 Trans-resistance amplifier for optical receivers

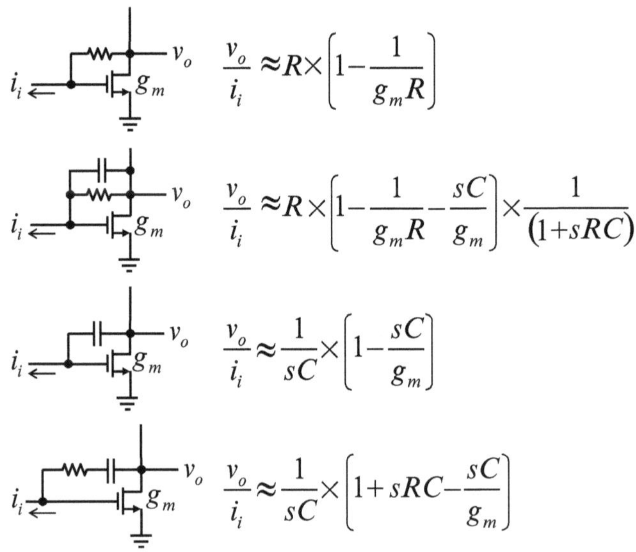

Fig. 2.33 Trans-impedances of four local shunt feedbacks

There are four possible cases of local shunt feedbacks with R and C as shown in Fig. 2.33. The first one is the standard trans-resistance amplifier. The third one is a Miller integrator. Since the signal is fed forward through the feedback capacitor into the inverting output, it makes a right-half plane zero. The fourth one is a straightforward voltage sum, and the series resistance with a capacitor makes a left-half-plane zero. By setting the RC value to be C/g_{m} or higher the right-half plane zero can be canceled or moved to the left-half plane. The second one is a current sum of two paths, which gives a pole and a right half-plane zero. It is the most demanding task for analog designers to derive the frequency response of this two-pole amplifier. Driving the shunt-feedback using a high-impedance current source complicates the hand analysis as two poles interact as shown in Fig. 2.34.

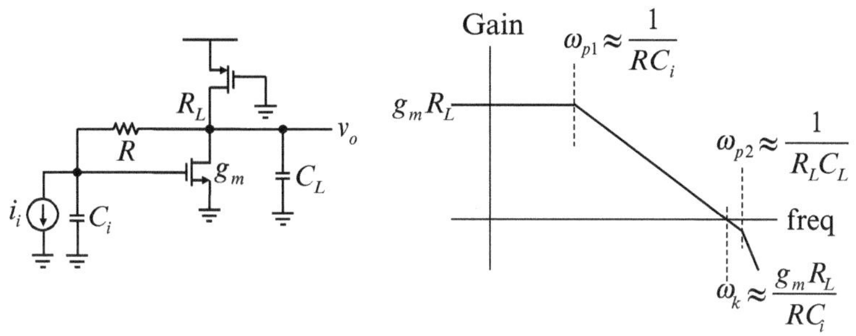

Fig. 2.34 Trans-resistance amplifier and open-loop gain

There are parasitic capacitances at the input and the output nodes marked as C_i and C_L ignoring the feedforward capacitance. The open-loop gain can be considered to analyze stability as follows.

$$\begin{aligned} a(s)f &= \frac{g_m}{\frac{1}{R_L} + sC_L + \frac{sC_i}{1+sRC_i}} \times \frac{1}{1+sRC_i} \\ &= \frac{g_m R_L}{1 + s\{RC_i + R_L(C_i + C_L)\} + s^2 RR_L C_i C_L}. \end{aligned} \tag{2.33}$$

This quadratic equation from the denominator cannot be factored algebraically. If $R \gg R_L$ is true, the $R_L C_i$ term can be ignored, and two factored poles become negative real. That is, they can be separated by more than the DC loop gain $g_m R_L$ for stability as explained in Fig. 2.34. Otherwise, two poles become complex conjugate poles on a circle with a radius equivalent to the geometric mean.

$$\omega_o = \sqrt{\omega_{p1}\omega_{p2}} \approx \sqrt{\frac{1}{RC_i} \times \frac{1}{R_L C_L}}. \tag{2.34}$$

To stay stable, the feedback loop pole frequency of $1/RC_i$ should be much lower than the output pole frequency of $1/R_L C_L$, which is often limited by the speed of the process technology.

Using (2.33), the closed-loop transfer function can be also derived as follows.

$$\frac{v_o}{i_i}(s) = \frac{g_m RR_L - R_L}{1 + g_m R_L + s\{RC_i + R_L(C_i + C_L)\} + s^2 RR_L C_i C_L}. \tag{2.35}$$

Now the closed-loop poles become even higher-Q poles as they move vertically further into the complex plane. They are on a circle with a radius of

$$\omega_o \approx \sqrt{(1 + g_m R_L) \times \frac{1}{RC_i} \times \frac{1}{R_L C_L}}. \tag{2.36}$$

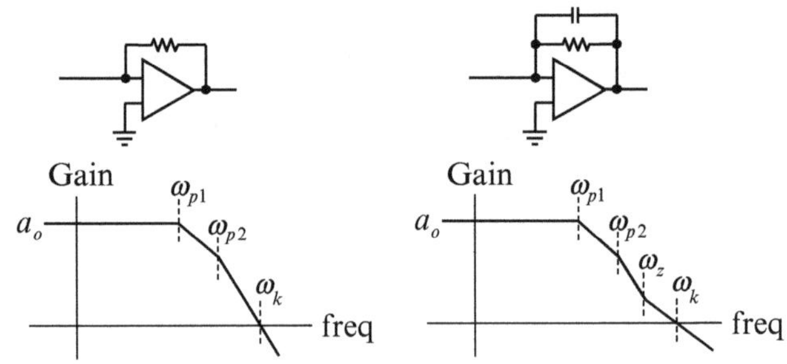

Fig. 2.35 Shunt feedbacks without and with a zero

The stability condition of two-pole networks can be approximately defined as $\omega_k < \omega_{p2}$ in open loop for the PM to be greater than 45°. That is,

$$R_L C_L < \frac{RC_i}{g_m R_L}, \tag{2.37}$$

which is difficult to meet in most wideband amplifier designs. The desirable solution is to add a zero below the unity loop-gain frequency. In general, the extra loop delay in the feedback path should be cancelled with a real zero inserted before the unity loop-gain frequency as explained in Fig. 2.35.

If the shunt feedback resistor is bypassed by a capacitor, it makes a zero in the open-loop gain at $1/RC$. Then the open-loop and closed-loop gains of (2.33) and (2.35) are modified as follows including C.

$$a(s)f = \frac{g_m}{\dfrac{1}{R_L} + sC_L + \dfrac{sC_i(1+sRC)}{1+sR(C+C_i)}} \times \frac{1+sRC}{1+sR(C+C_i)}$$
$$= \frac{g_m R_L(1+sRC)}{1+s\{R(C+C_i)+R_L(C_i+C_L)\}+s^2 R R_L(CC_i+CC_L+C_iC_L)}. \tag{2.38}$$

$$\frac{v_o}{i_i}(s) = \frac{g_m R_L R\left(1-\dfrac{1}{g_m R}-\dfrac{sC}{g_m}\right)}{1+g_m R_L+s[R\{C_i+(1+g_m R_L)C\}+R_L(C_i+C_L)]+s^2 R R_L(CC_i+CC_L+C_iC_L)}.$$

These equations further complicate the assessment of stability with greater complexity, but one thing to note is that two poles are moving farther from the imaginary axis as the first-order term increases as the loop gain increases. It implies the Q of the poles gets lower, but the right half-plane zero is created in the close-loop gain due to feedforward through the capacitor C.

If $R \gg R_L$ is true again as before, the $R_L C_i$ term in the denominator can be ignored, and two factored poles in the open-loop gain become negative real as expected.

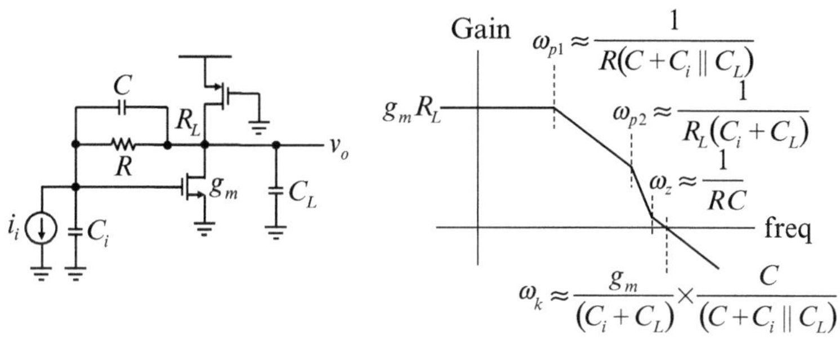

Fig. 2.36 Trans-resistance amplifier with shunt compensation and open-loop gain

$$\omega_{\mathrm{p1}} \approx \frac{1}{R(C + C_{\mathrm{i}} \| C_{\mathrm{L}})} \quad \text{and} \quad \omega_{\mathrm{p2}} \approx \frac{1}{R_{\mathrm{L}}(C_{\mathrm{i}} + C_{\mathrm{L}})}, \tag{2.39}$$

where $(C_{\mathrm{i}} \| C_{\mathrm{L}})$ denotes the value of $C_{\mathrm{i}}C_{\mathrm{L}}/(C_{\mathrm{i}} + C_{\mathrm{L}})$ for the series connection of two capacitors. The open-loop gain is shown in Fig. 2.36.

Again for the PM to be greater than 45° for this case, $\omega_z < \omega_k$ so that the zero frequency can be lower than the unity loop-gain frequency. That is,

$$RC > \frac{C + C_{\mathrm{i}} \| C_{\mathrm{L}}}{C} \times \frac{C_{\mathrm{i}} + C_{\mathrm{L}}}{g_{\mathrm{m}}}, \tag{2.40}$$

which can be easily met. Otherwise, two open-loop poles become complex conjugate high-Q poles on a circle with a radius of

$$\omega_{\mathrm{o}} = \sqrt{\omega_{\mathrm{p1}}\omega_{\mathrm{p2}}} \approx \sqrt{\frac{1}{R(C + C_{\mathrm{i}} \| C_{\mathrm{L}})} \times \frac{1}{R_{\mathrm{L}}(C_{\mathrm{i}} + C_{\mathrm{L}})}}. \tag{2.41}$$

Similarly, the closed-loop poles would move to a new circle but with a lower Q since the zero pulls the poles away from the imaginary axis.

$$\omega_{\mathrm{o}} \approx \sqrt{(1 + g_{\mathrm{m}}R_{\mathrm{L}}) \times \frac{1}{R(C + C_{\mathrm{i}} \| C_{\mathrm{L}})} \times \frac{1}{R_{\mathrm{L}}(C_{\mathrm{i}} + C_{\mathrm{L}})}}. \tag{2.42}$$

This is the same result as obtained by the famous pole-splitting Miller effect of a two-pole system. If the Miller capacitance C is very large and there is no shunt feedback, the dominant pole is generated by the Miller capacitance at the input, and the non-dominant pole is created by the sum of input and output capacitances driven by the trans-conductance. From (2.38), two poles can be derived as follows using the dominant pole approximation and the geometric mean.

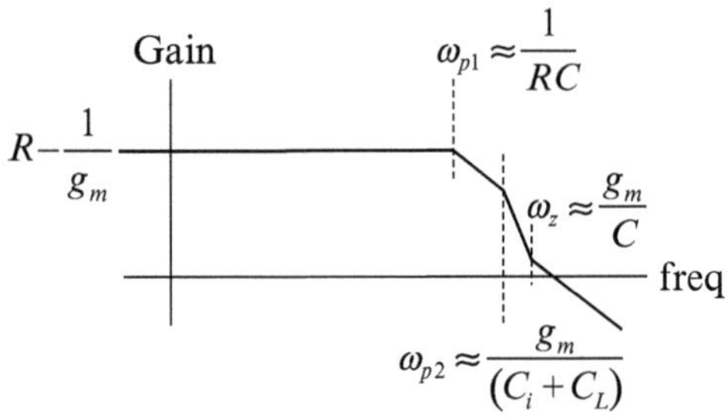

Fig. 2.37 Trans-resistance amplifier gain with two widely separated poles

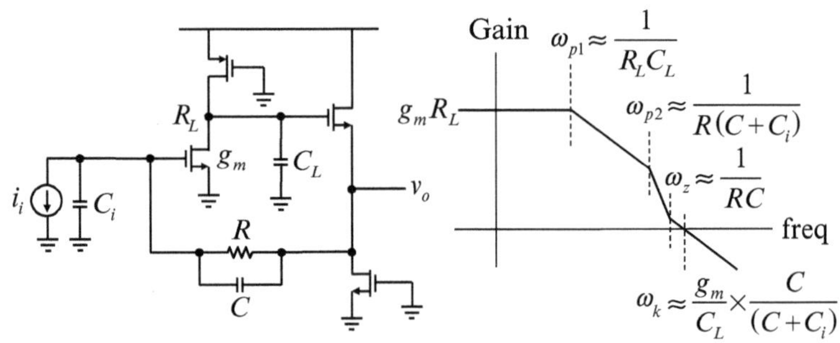

Fig. 2.38 Buffered trans-resistance amplifier and open-loop gain

$$\begin{aligned}\omega_{\mathrm{p1}} &\approx \frac{1 + g_{\mathrm{m}} R_{\mathrm{L}}}{R\{C_{\mathrm{i}} + (1 + g_{\mathrm{m}} R_{\mathrm{L}})C\}} \approx \frac{1}{RC}, \\ \omega_{\mathrm{p2}} &\approx \frac{C_{\mathrm{i}} + (1 + g_{\mathrm{m}} R_{\mathrm{L}})C}{R_{\mathrm{L}}(CC_{\mathrm{i}} + CC_{\mathrm{L}} + C_{\mathrm{i}} C_{\mathrm{L}})} \approx \frac{g_{\mathrm{m}}}{C_{\mathrm{i}} + C_{\mathrm{L}}}.\end{aligned} \tag{2.43}$$

The only difference in this shunt feedback is that the Miller pole is now at $-1/RC$ due to the shunt-feedback resistor. The frequency response of the trans-resistance amplifier with two widely separated real open-loop poles is sketched in Fig. 2.37.

The trans-resistance amplifier with feedforward compensation offers a desirable very high-frequency dominant pole, and the input and output capacitances are driven by the diode resistance of $1/g_{\mathrm{m}}$. It is because the feedforward path goes through the shunt capacitance. One way to eliminate the right half-plane zero and make the non-dominant second pole independent of the input capacitance is to use a feedback buffer amplifier. The low impedance of the buffer amplifier output can stop the signal feedforward, but the feedback path is not affected. Therefore, there is no right half-plane zero created, and the trans-conductance doesn't need to drive the input capacitance.

The buffered trans-resistance amplifier is shown along with its open-loop gain in Fig. 2.38. In most multi-stage wideband amplifier designs, it is common to use a source-follower buffer for Miller capacitance feedback. Assuming that the source follower has an ideal unity gain and the pole at the source follower output is high enough to ignore, the open- and closed-loop gains can be obtained as follows.

$$a(s)f = \frac{g_m R_L(1+sRC)}{(1+sR_L C_L)\{1+sR(C+C_i)\}}.$$

$$\frac{v_o}{i_i}(s) = \frac{g_m R_L R}{1+g_m R_L + s[R\{C_i+(1+g_m R_L)C\}+R_L C_L]+s^2 R_L R(C+C_i)C_L}. \tag{2.44}$$

Note that there is no right half-plane zero, and both the open-loop gain and the closed-loop gain are greatly simplified.

If $R \gg R_L$ is true, two factored poles are

$$\omega_{p1} \approx \frac{1}{R_L C_L} \quad \text{and} \quad \omega_{p2} \approx \frac{1}{R(C+C_i)}, \tag{2.45}$$

respectively, as shown in Fig. 2.38. For PM to be greater than 45°,

$$g_m R > \frac{C_L}{C} \times \frac{(C+C_i)}{C}, \tag{2.46}$$

which can be easily met. The closed-loop poles would move to a new circle but with a lower Q since the zero pulls the poles away from the imaginary axis.

$$\omega_o \approx \sqrt{(1+g_m R_L) \times \frac{1}{R(C+C_i)} \times \frac{1}{R_L C_L}}. \tag{2.47}$$

If the Miller capacitance C is very large, two familiar widely separated poles can be approximated as follows.

$$\begin{aligned} \omega_{p1} &\approx \frac{1+g_m R_L}{R\{C_i+(1+g_m R_L)C\}} \approx \frac{1}{RC}, \\ \omega_{p2} &\approx \frac{C_i+(1+g_m R_L)C}{R_L(C+C_i)C_L} \approx \frac{g_m}{C_L}. \end{aligned} \tag{2.48}$$

The frequency response of this case is shown in Fig. 2.39.

The shunt feedback implements wideband amplifiers with both low input and output impedances, but makes an extra pole, thereby requiring frequency compensation for stability. It is possible to further broadband the shunt feedback amplifier with dual or triple gain stages as shown in Fig. 2.40.

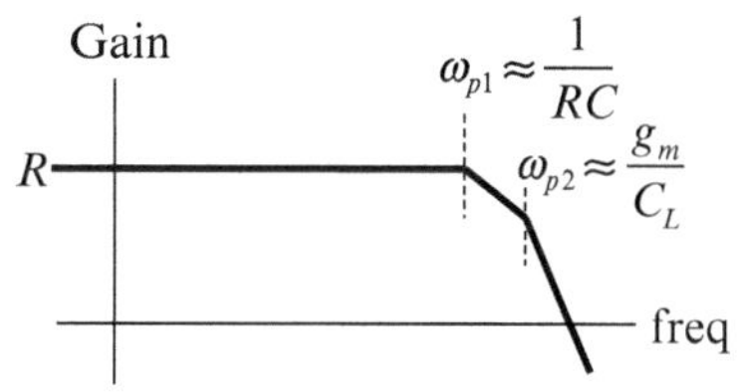

Fig. 2.39 Buffered trans-resistance amplifier gain with two widely separated poles

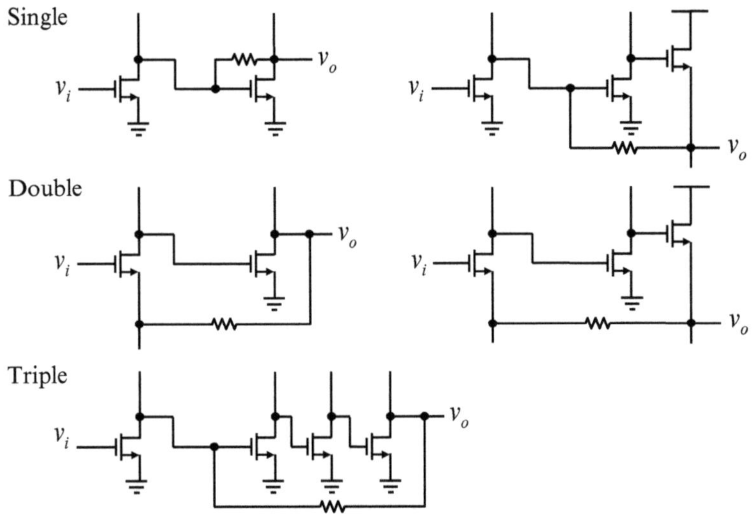

Fig 2.40 Trans-resistance amplifiers with single, double, and triple gain stages

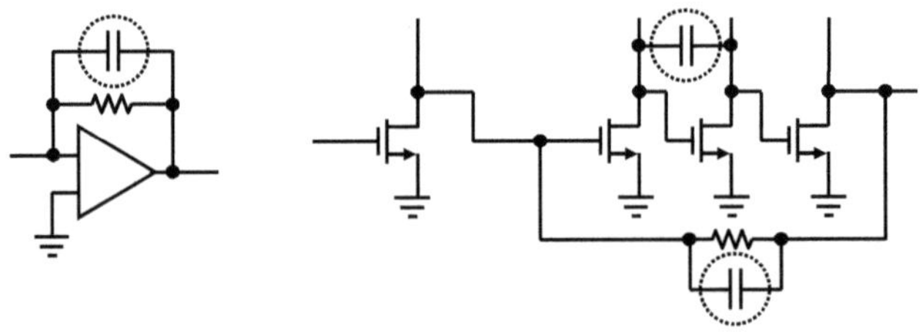

Fig. 2.41 Feedforward frequency compensations

However, it is a formidable task to stabilize three- or four-pole response though the buffered feedback helps to reduce the feedforward effect. There is no simple way, but the common wisdom is to insert as many feedforward zeros that bypass gain stages like using the Miller capacitance. Since zeros should be added after poles, the gain attainable by extra poles is limited and incremental. Stabilizing the loop with multiple integrators in $\Delta\Sigma$ modulators is a good example of the feedforward compensation.

Examples of the feedforward compensation are shown in Fig. 2.41. It is to lower the path impedance between the two nodes, and to let the signal directly pass through the capacitor at frequencies higher than $1/RC$ or g_m/C, which creates a zero effect by definition. If it bypasses the inverting signal, the zero moves to the right half plane. There are three ways to cancel the right half-plane zero. The source follower feedback cuts the feedforward path, but its pole in the feedback loop creates left-half plane zero. The G_m boosting moves the zero to higher frequencies, but extra pole inside the local feedback loop for G_m boosting complicates the overall settling. Lastly, the right half-plane zero is canceled and moved to the left half plane by just adding a resistance in series with the capacitor.

2.4 G_m Boosting and Noise Cancellation

A need arises to make an effective trans-conductance larger than it actually is for buffering and low noise. Active shunt feedback either lowers the resistance level, or boosts the conductance level as shown in Fig. 2.42.

When looking into the input side, the shunt feedback resistance looks smaller by the amplifier gain. By active shunt feedback, the input conductance is made very small. If the output is taken from the shunt transistor, this common-source (CS) and common-drain (CD) stages can be used to boost the output resistance of the series feedback as shown in Fig. 2.43.

It also shows a G_m-boosted source follower with CS and common-gate (CG) feedback. The former raises the output resistance by the loop gain as sketched

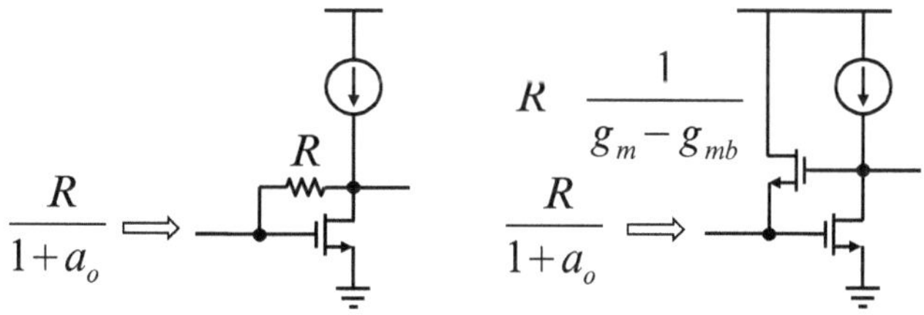

Fig. 2.42 Active shunt-feedback by G_m boosting

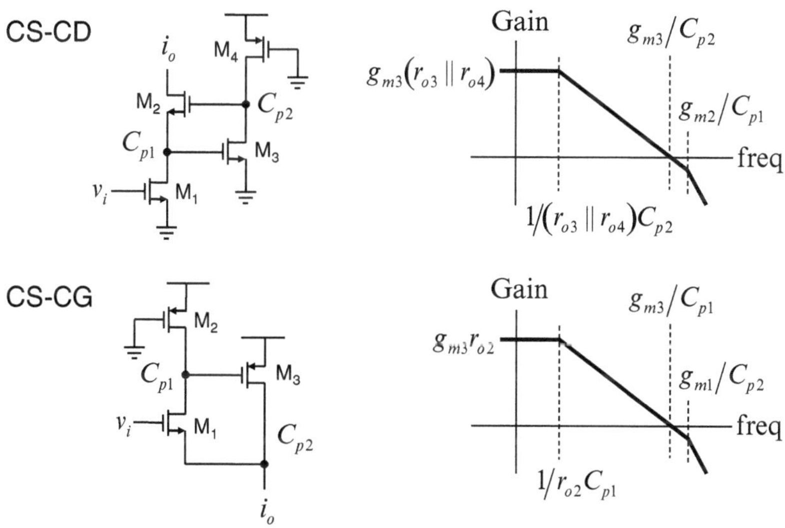

Fig. 2.43 Two examples of active G_m boosting

while the latter lowers the output resistance by the loop gain. Boosted transconductance helps to lower impedance and widen bandwidth. That is, their transconductances are further increased to be

$$(g_{m2} - g_{mb2}) \times g_{m3}(r_{o3} \| r_{o4}) \quad \text{and} \quad g_{m1} \times g_{m3} r_{o2}, \tag{2.49}$$

respectively. Due to these shunt feedback gains, the gain-boosted stage and the super-G_m source follower are made practical overcoming the handicap of low transconductance values of MOS transistors.

Examples of two super-G_m source followers are shown in Fig. 2.44. Due to the body effect, their gains approach the same $g_{m1}/(g_{m1} - g_{mb1})$ as the regular source follower, but their output resistances are made much lower enhancing the load drive capacity greatly.

Noise sources in feedback circuits are shown in Fig. 2.45. Feedback only enhances analog performance limited by deterministic parameters, but noise is a random power with a variance with no magnitude information. Therefore, all noise powers in feedback networks are added without being lowered by the negative feedback. Noise is further enhanced in high-Q circuits like resonators.

The strategies to achieve low noise in open-loop LNA are mostly of two kinds. One is to make the effective G_m higher than real G_m, which contributes to actual noise, and the other is noise cancellation. Narrow-banding is another way, but system requirements set the bandwidth. Oversampling lowers the in-band noise, but pays the speed penalty.

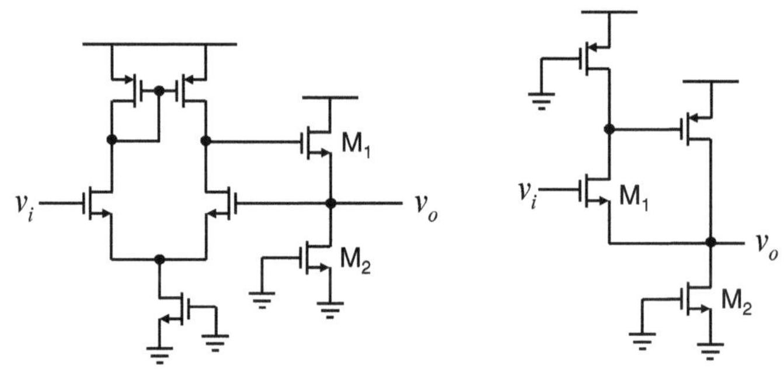

Fig. 2.44 Two G_m-boosted source follower examples

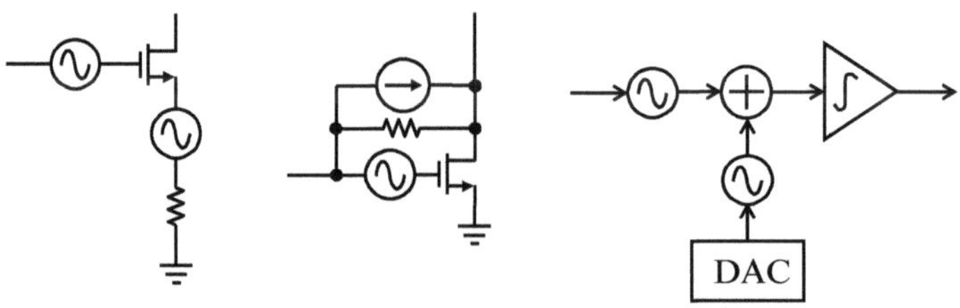

Fig. 2.45 Two G_m-boosted source follower examples

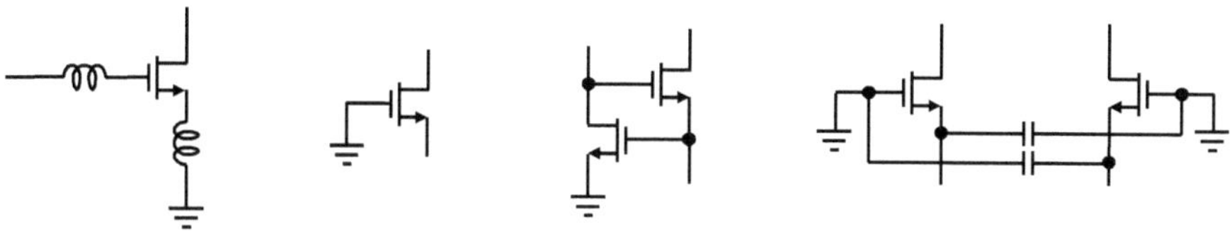

Fig. 2.46 Low-noise amplifier schemes

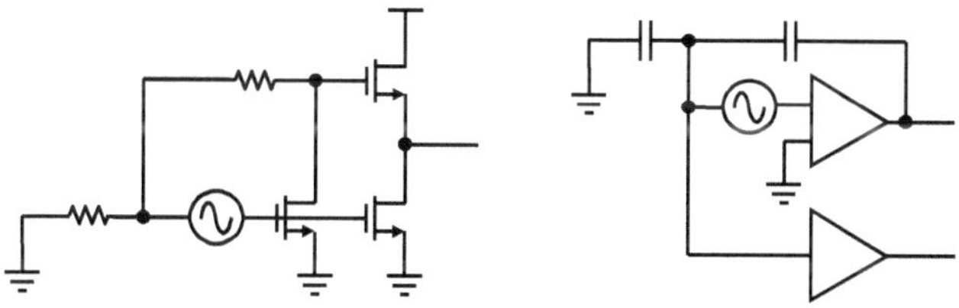

Fig. 2.47 Concept of noise cancellation by feedforward

Widely used low-noise techniques that increase the effective G_m are shown in Fig. 2.46. The passive inductor degeneration is very effective to achieve low noise. The CG amplifier has a factor of 2/3, but the empirical filling factor doesn't justify its effectiveness. The G_m boosting enhances the effective G_m, but the feedback amplifier contributes some noise. Lastly, multiplying the signal by feeding forward through capacitors doubles the input swing, but capacitors get larger and extra power is demanded. If G_m enhancement is by adding gain stages, they also add noise and power. There are no obvious solutions to LNA designs, but new process improves the noise performance incrementally.

Alternatively, feedforward can cancel the in-band noise as sketched in Fig. 2.47. It is a two-path system for noise [2, 3]. Although the noise polarity is not known, one noise source can be amplified through two identical gain paths with inverting and non-inverting gains, and summed later. The end result of this summing is the cancellation of the in-band noise of one source. If two-path gains are matched, everything is cancelled, but noise and offset of the cancelling path remain. That is, out-of-band noise and the noise of the additional amplifier are not cancelled. When designing such noise-cancelling amplifiers, the difficulty also lies in achieving large signal linearity.

References

1. B. Song, *MicroCMOS Design* (CRC, Boca Raton, 2012), pp. 32–37
2. F. Bruccoleri, E. Klumperink, B. Nauta, Noise cancelling in wideband CMOS LNAs, *ISSCC Dig. Tech. Papers* (Feb. 2002), pp. 406–407
3. Y. Miyahara, M. Sano, K. Koyama, T. Suzuki, K. Hamashita, B. Song, A 14b 60MS/s pipelined ADC adaptively cancelling opamp gain and nonlinearity. IEEE J. Solid State Circuits **50**, 416–425 (2014)

Chapter 3
Almost DC Circuits

There are two types of circuits operating at almost DC. They are power supplies and sensor networks, in particular human body sensors for medical instrumentations. The former requires very low output impedance for transient uses while the latter needs narrow bandwidth. All electronic circuits need to draw energy from constant DC power sources for proper operations. Power supplies perform two basic functions of DC–DC conversion and voltage regulation. Voltage regulator is to reduce ripple in the supply output, but consumes power due to the voltage drop in the pass transistor. In recent years, switching DC–DC converters are commonly used for high efficiency together with low-dropout (LDO) regulators. Almost DC sensing circuits are for high-impedance instrumentations such as voltmeters, which suffer classic problems such as DC wander and motion artifact due to capacitive coupling.

3.1 Regulator DC Performance

The main function of the power supply is to take an unregulated supply voltage from the source such as a rectifier and to provide a constant DC voltage to the varying load. Two examples of the class-A voltage regulator are shown together with the regulator concept in Fig. 3.1.

Regulator continuously adjusts the load current so that the loaded output voltage can stay constant as sketched at the top. This voltage-controlled current source can be made using either a pass transistor or a variable resistor. The former is the standard class-A voltage regulator, and the latter is a low-dropout (LDO) version commonly operating the regulator with low voltage drop. The MOS transistor biased in triode region can be also used as a variable resistor.

The DC performance of the regulator is measured by the load regulation and power efficiency. The load regulation is defined as the % voltage drop at the peak

B.-S. Song, *System-level Techniques for Analog Performance Enhancement*,
DOI 10.1007/978-3-319-27921-3_3

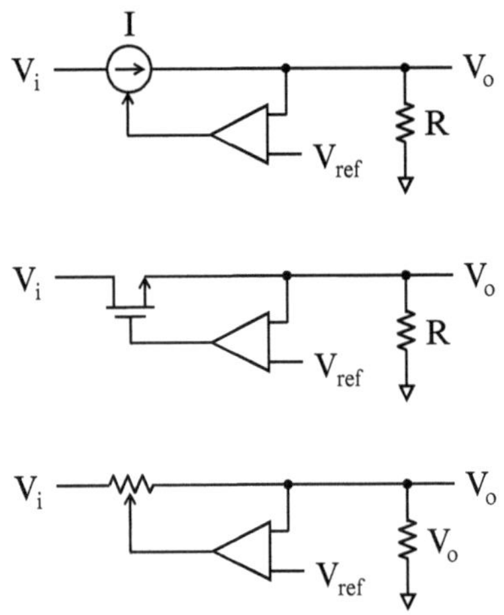

Fig. 3.1 Two implementations of voltage regulators

load current. The smaller the value is, the better the regulator is. The error amplifier is usually made of a feedback opamp, and its open-loop driving point output resistance is lowered by the DC open-loop gain.

$$\text{Load regulation} = \frac{\Delta V_{\text{o}}}{V_{\text{o}}} = \frac{I_{\max} \times R_{\text{o}}}{V_{\text{o}}} \approx \frac{I_{\max} \times \frac{r_{\text{o}}}{a_{\text{o}}}}{V_{\text{o}}}, \tag{3.1}$$

where r_o and a_o are the open-loop output resistance and the DC loop gain of the regulator, respectively. The closed-loop output resistance R_o of the regulator is reduced by the shunt-feedback loop gain a_o. This implies that the regulated output stays constant for any changes in loading since the voltage drop is regulated to keep it constant by high-gain feedback.

Since the same current flows through the regulator and the load, the power efficiency can be defined just as the voltage ratio.

$$\eta = \frac{V_{\text{o}}}{V_{\text{i}}} = 1 - \frac{V_{\text{i}} - V_{\text{o}}}{V_{\text{i}}}. \tag{3.2}$$

From (3.2), note that the power efficiency drops as the voltage drop across the regulator increases. For higher efficiency, it is mandatory to minimize the voltage drop in the regulator, which is the basic LDO concept. Therefore, the regulator can be divided into two separate functional blocks. One is the DC–DC converter, and the other is the LDO which is still the standard class-A regulator. Most common on-chip regulators consist of narrowband opamps that supply the average current to the load and large external storage capacitors as shown at the top on Fig. 3.2.

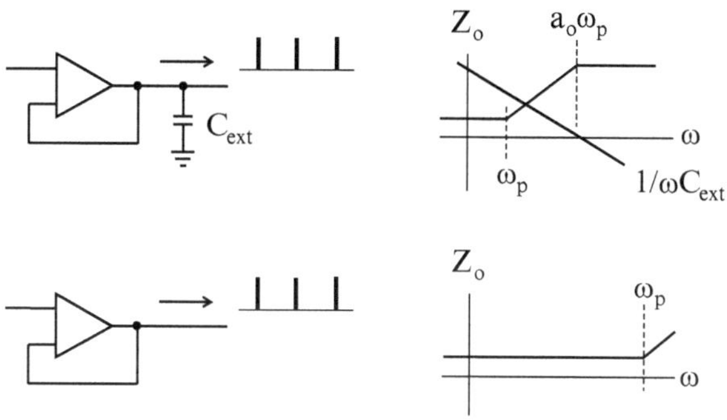

Fig. 3.2 Two regulator versions with narrowband and wideband opamps

3.2 Regulator Transient Performance

The transient performance of the regulator or LDO is measured by the residual ripple and glitch present in the regulated output. The whole purpose of adding the external capacitor C_{ext} is to suppress the voltage dips resulting from the current spikes that are common in switching digital power supplies. The output impedance Z_o of the opamp becomes higher as the loop gain decreases above the dominant pole frequency ω_p, and stays high after the unity loop gain frequency of $a_o\omega_p$. That is, the slow high-impedance opamp supplies only the average current to the load while the external capacitor C_{ext} absorbs the sudden surge of the transient current since the capacitor impedance is lower at high frequencies. The parallel sum of the output impedances of the opamp and the external capacitor can be kept low at high frequencies. Capacitor-free regulators or LDOs are often tried for the low bill of materials [1, 2]. They require wideband opamps to keep the output impedances very low even at high frequencies as sketched at the bottom of Fig. 3.2. It is impractical to replace the large external bypass capacitor with a wideband opamp.

As digital circuits are clocked at GHz rates, the current spikes due to the high current demand in digital supplies last only for a fraction of ns. The capacitor-free regulator is too slow to respond to them, and only the large C_{ext} can effectively suppress the current spike. It is shown that the external capacitor integrates the current spike in the form of the gate function, but the high-impedance slow opamp recovers from it slowly with its own time constant as shown in Fig. 3.3.

In the former case, the ripple voltage is reduced to be small as the current spikes are integrated. However, in the latter case, both the current and voltage glitches remain large as the disturbed opamp output tries to settle back. This implies that transients from the load side can be effectively suppressed only using bypass capacitors while any ripples from the source side can be filtered out by narrowband feedback opamps.

Power supply is to reduce ripple both from the input and output sides. The former is by the line regulation while the latter is by the load regulation. In addition, among mixed-mode analog designers, there has been a golden rule not to break: "Do not mix supply currents." In mixed-mode chips, power supplies to most

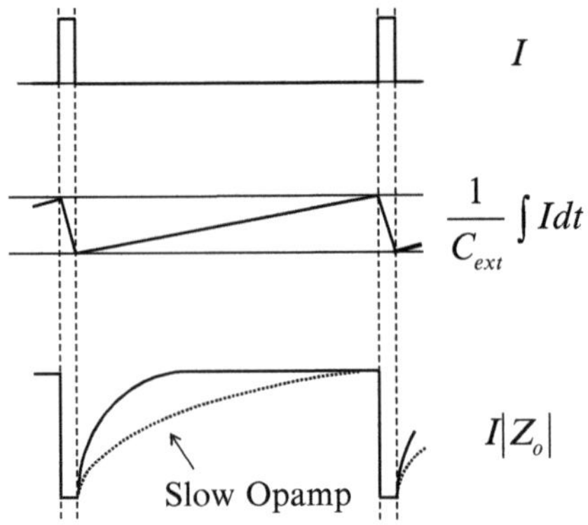

Fig. 3.3 Current spike, integrated by C_{ext}, and capacitor-free regulator output

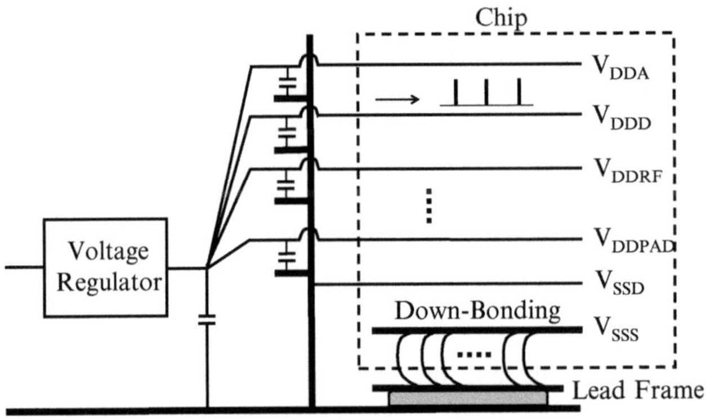

Fig. 3.4 Power supplies and grounding plan for mixed-mode chips

functional blocks such as analog, digital, LNA, mixer, VCO, pad, analog and digital substrates are physically separated, and their currents are never mixed to avoid coupling among them. It is mainly to prevent the switching circuits such as digital and mixer from being coupled to the sensitive circuits such as LNA and VCO. What has been commonly called "substrate coupling" results from the substrate potential difference due to poor grounding. Figure 3.4 illustrates the supply and grounding scheme for mixed-mode analog/digital/RF chips.

The solid line is the physically solid ground plane. All chip grounds and substrates are directly multiple bonded to the lead frame, which is also glued conductively to the PCB ground. Flip-chip packaging can reduce bonding wire inductance. The digital ground can be either downbonded to the lead frame together with all other analog grounds, or taken out to ground it to the PCB digital ground plane.

A voltage regulator with a large external capacitor sits on a star ground point, from where all power lines can be routed. Considering long power lines on the PCB plus bonding wires, the supply lines commonly suffer from large voltage ripples and glitches mainly resulting from the voltage drop of $L(di/dt)$. Note that the chip grounds are heavily wire-bonded to the PCB ground. To kill the ripples and glitches on the power lines, each supply pinout of the chip should be decoupled using bypass capacitors at the closest point to each pin. Digital supplies can be decoupled to the digital ground plane on the PCB, and both analog and digital ground planes are connected at the board level to one optimal middle point. This arrangement effectively separates all supply lines, and prevents supply currents from being mixed.

3.3 Lossless Energy Transfer

All operations in electronics consume energy as is true in the nature. With no energy consumed, everything stays unchanged. If any change occurs, it implies that the energy of the changed state is lower than that of the previous state since some energy should be lost in the process. The only case of the lossless energy transfer between two circuit elements is the *LC* resonance as shown in Fig. 3.5. The capacitor *C* (Farad) and the inductor *L* (Henry) are only two energy storing elements holding the energy in the form of the charge *Q* (Coulomb) and the flux ψ (Weber), respectively. The current is integrated to give the voltage in the capacitor while the voltage is integrated to give the current in the inductor. They are dual in concept. If *V* and *I* are flipped, their charge and flux relations are related with the same equations.

The unit of energy is Joule. When the capacitor voltage is *V* and the inductor current is *I*, the energies they are holding are

$$E = \int_0^V Q dV = C\int_0^V V dV = \frac{CV^2}{2}, \tag{3.3}$$

and

$$E = \int_0^I \psi \, dI = L\int_0^I I \, dI = \frac{LI^2}{2}, \tag{3.4}$$

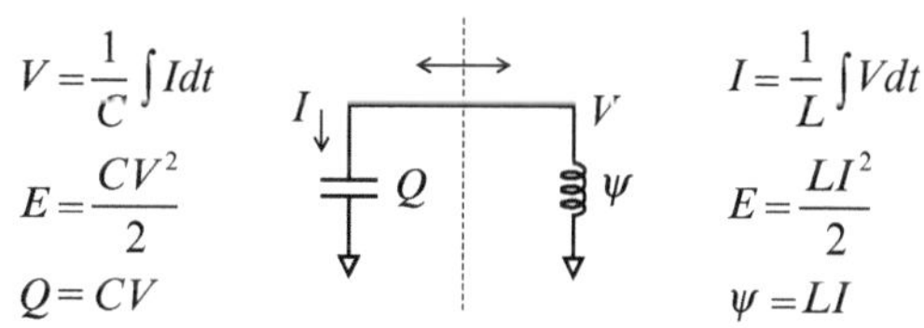

Fig. 3.5 Only lossless energy transfer between inductor and capacitor

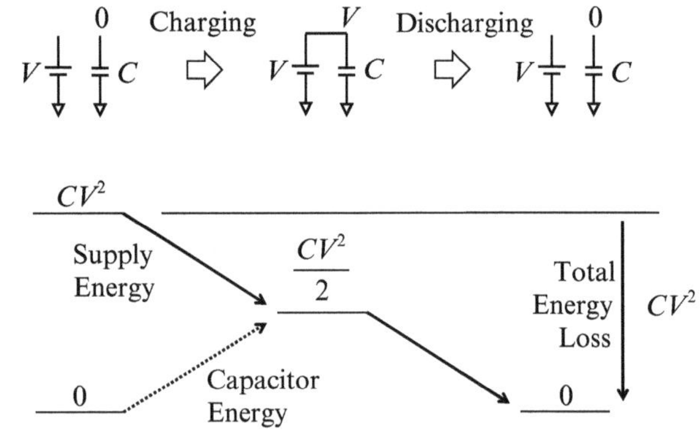

Fig. 3.6 Energy loss for charging and discharging C

respectively. The resonant tank circuit is analogous to the microwave resonant cavity, which can hold energy inside. The capacitor voltage is integrated to give the inductor current, and the inductor current is integrated back to give the capacitor voltage. This two 90° delays in the integration makes the 180° total delay that leads to the resonance. Without a loss in the inductor or capacitor, the resonant energy stays constant while it is transformed into the charge or flux form alternately.

That is, storing current on inductor in a flux form is lossless if the inductor voltage is switched. Similarly, the capacitor can be charged using current without energy loss. However, current source is a unipolar element, and there should be a voltage drop across any current source. Therefore, charging capacitor using voltage source is the only option that consumes energy. The change in the energy level involved in one event of charging and discharging a capacitor C is explained in Fig. 3.6.

At the top row, the capacitor holding 0 V is shown to be charged up by the voltage source V, and discharged back to 0 V. At the bottom row, it is shown that the energy level of the capacitor is raised from 0 to $CV^2/2$, and moved back to 0. The charging energy $CV^2/2$, which is the same as the stored energy, comes from the source. Therefore, one event of charging C to V and discharging it to 0 consumes the total energy of CV^2. As in the digital circuits clocked at the rate of f, if this charging and discharging event happens f times per second, the energy loss of CV^2 occurs f times per every second. Thus the power consumption with a unit of Watt is

$$P = fCV^2. \tag{3.5}$$

Note that in this energy transfer using capacitor, only $CV^2/2$ is delivered to the load, and the power efficiency is lower than 50 % at the most. That is, there is no energy transfer to capacitors possible without consuming energy. If energy is transferred using capacitors, the capacitor charging energy is lost, and only the stored energy which is equal to the charging energy is delivered to the load [3, 4].

The exhaustive search leads only to the four arrangements for possible energy storage as shown in Fig. 3.7. The basic function of the power supply is to take as much energy as needed from the source, and to deliver it to the load while keeping

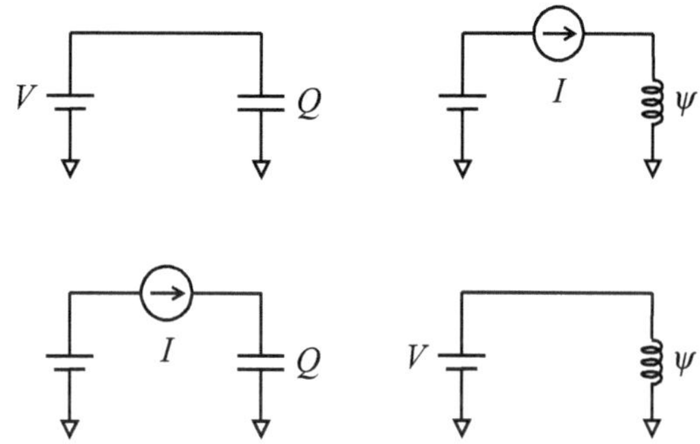

Fig. 3.7 Four ways to store energy

the output voltage constant. The source can be either a voltage source or a current source while the intermediate energy storing element can be either a capacitor or an inductor. As in the lossless *LC* resonant case, only the two cases of the bottom row can be lossless since the capacitor and the inductor are charged with the current and the voltage, respectively. The two cases at the top row are automatically disqualified as the energy transfer is lossy.

Even from the two cases at the bottom row, charging the capacitor with the current source is also lossy since the current source is unipolar by nature. The current only flows from the higher voltage side to the lower side. Unlike the *LC* resonant case that the inductor current charges the capacitor voltage, current source consumes energy in reality to charge the capacitor due to the voltage drop in the current source. Therefore, the only real lossless intermediate energy storage scheme is to charge the inductor with voltage, which is widely used in all high-efficiency switching regulator for modern power-management chips.

3.4 Switched-Capacitor DC–DC Converters

Although half the energy is lost in the energy transfer if the capacitive energy storage is used, switched-capacitor DC–DC up/down converters can be implemented handily in particular for low-power uses as no bulky inductors are used [5, 6]. The switched-capacitor up-conversion is common in all on-chip clock boost circuits with low supplies. Its example with an LDO is conceptually sketched in Fig. 3.8, where the input DC V_i is down-converted to V_i^*.

Then V_i^* is regulated by the LDO to yield the output V_o. The storage capacitor C^* is a small capacitor to hold V_i^*, but cannot provide enough ripple filtering. If it is made large, the ripple will be reduced more, but the DC–DC converter will behave sluggishly. The LDO is more effective in removing the ripple and glitch. The reference voltages V_{ref1} and V_{ref2} are to get V_i^* and V_o, respectively.

Two feedback loops are applied. The first one is the discrete-time servo feedback to control the capacitor switches, and the latter is the continuous-time feedback

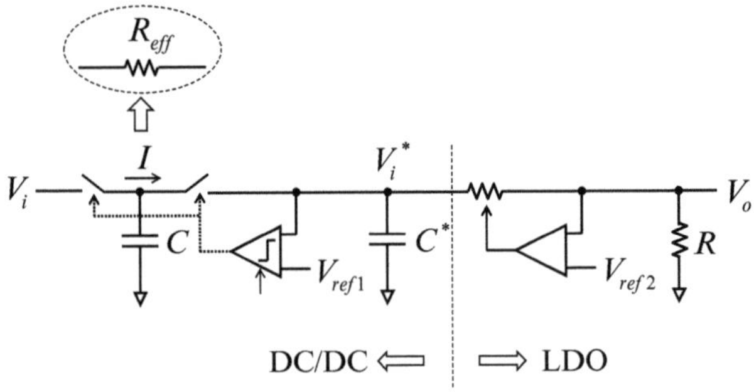

Fig. 3.8 Switched-capacitor down-converter with LDO

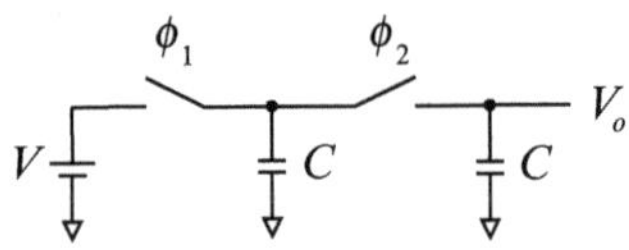

Fig. 3.9 Switched-capacitor charge-redistribution voltage divider

using an error opamp. The former is based on the pulse-density modulation like the standard $\Delta\Sigma$ modulator. The comparator gets the 1b error polarity for the Δ function while the capacitor C^* integrates the error charge to perform the Σ function. This oversampled feedback adjusts the clock rate so that the following condition is met.

$$R_{\mathrm{eff}} = \frac{1}{fC} = \frac{V_{\mathrm{i}} - V_{\mathrm{i}}^*}{V_{\mathrm{o}}} \times R, \tag{3.6}$$

where R_{eff} is the effective equivalent resistance of the capacitor C switched at the clock rate of f. Since the same current flows from the input to the output, the power efficiency becomes the voltage ratio of the output to the input as in the class-A regulator.

$$\eta = \frac{V_{\mathrm{o}}}{V_{\mathrm{i}}}. \tag{3.7}$$

Even the charge redistribution consumes energy though the total charge is conserved. If switches are ideal, the input voltage is sampled on C during one clock phase ϕ_1, and the sampled charge $Q = CV$ is redistributed with another C during the other clock phase ϕ_2 as shown in Fig. 3.9. However, if energy stays the same, the charge cannot be redistributed as explained in Fig. 3.10.

The energy states during two clock phases differ. Like the energy used to charge the capacitor, one capacitor loses energy while the other capacitor gains the same

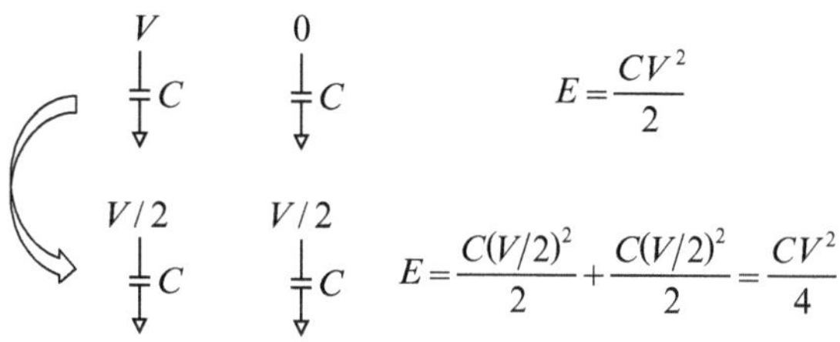

Fig. 3.10 Energy loss for charge redistribution

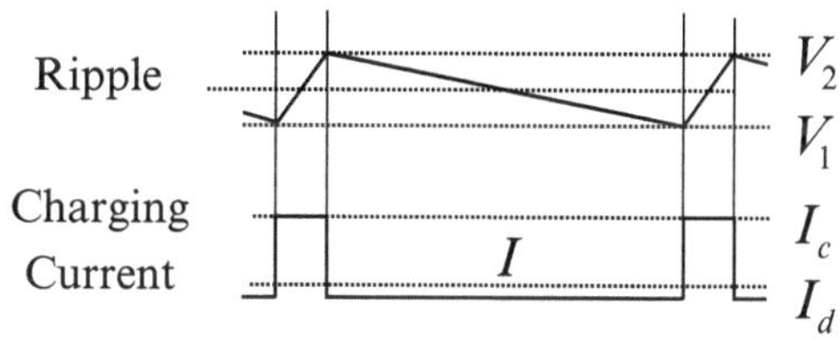

Fig. 3.11 Ripple causes charging and discharging

energy. It is like the water in one full glass can fill two glasses half full. Due to this energy loss, once redistributed, it will never go back to the previous state during phase ϕ_1.

The cold boot is to start charging the energy storing element from the initial no-energy state. Once filled up, the charging and discharging of the storage repeat as the load draws current from the storage, and the ripple is created as shown in Fig. 3.11, where the load current is assumed to be constant I. The high and low voltages of the ripple V_2 and V_1 are close, and the magnitude of the ripple is given by the current, the storage capacitor, and the charging/discharging time.

$$V_2 - V_1 = \left(\frac{I_c - I}{C^*}\right) \times t_c = \left(\frac{I - I_d}{C^*}\right) \times t_d, \tag{3.8}$$

where the subscript c and d denote charging and discharging, respectively. From (3.8), the average current supplied to the load can be estimated as

$$I = \frac{I_c t_c + I_d t_d}{t_c + t_d}. \tag{3.9}$$

The ripple can be reduced using a large capacitor or clocking fast. No matter how small the ripple is, the charging current should raise the stored energy level so that the load can draw the same energy from the storage. The same energy relation for charging and discharging can be applied to the incremental ripple energy of the storage capacitor.

$$E = \frac{1}{2}C(V_2 - V_1)^2 = C\left(\frac{V_1 + V_2}{2}\right)(V_2 - V_1) = \frac{1}{2}CV_2^2 \quad \text{if } V_1 = 0. \tag{3.10}$$

Unlike the lossy switched-capacitor DC–DC down-conversion, there is no charge redistribution, and there is no energy loss in the up-conversion itself. However, the power efficiency is still limited to the same 50 % since the incremental ripple energy should be supplied as given by (3.10). There have been several efforts made in recent works of switched-capacitor power supply. One is that if the ripple is small, there is no energy consumed in charging and discharging. The other is that the energy of a small capacitor can be temporarily transferred to a larger capacitor, and later the stored energy can be recycled to charge back the small capacitor [7–9].

3.5 Switched-Inductor DC–DC Converters

As implied by (3.9), the principle of switching DC–DC converters is to supply pulse-modulated current to the load, and the switched current is averaged in the intermediate energy storing element. The residual ripple and glitch are further eliminated in the later stage using the LDO and the external capacitor. The switched-inductor DC–DC converter can be configured as shown in Fig. 3.12.

The buck and boost converters (down- and up-converters) can be made by switching the inductor input and the output, respectively as shown. As discussed with Fig. 3.7, the only way to store energy with no energy loss is by charging the inductor current using a voltage source, and the remaining three cases of charging the energy storing elements are lossy. Therefore, if the switch-on-resistance is minimized and high-Q inductors are used, very high-power efficiency is attainable in this DC–DC conversion.

The same discrete-time servo feedback compares the output with V_{ref}, and feeds either the pulse-width or pulse-density modulated error voltage back so that the inductor current can be charged to meet the average current demand from the load side. The pulse-width modulation has been used in power supply boards populated by discrete components, but the pulse-density modulation as used in the ΔΣ modulator is simpler to implement in integrated power-management chips. The

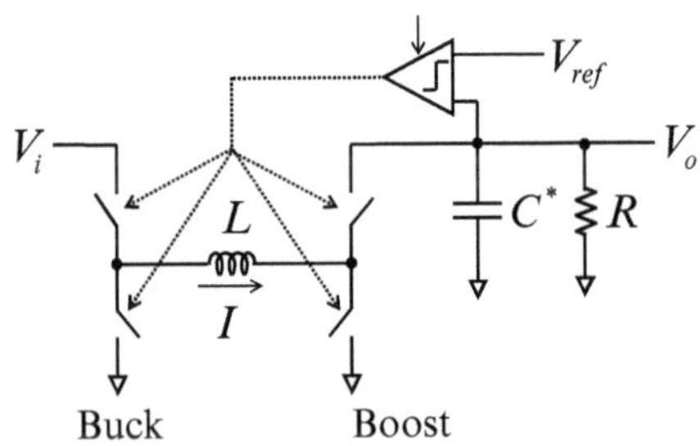

Fig. 3.12 Switched-inductor DC–DC buck-and-boost converters

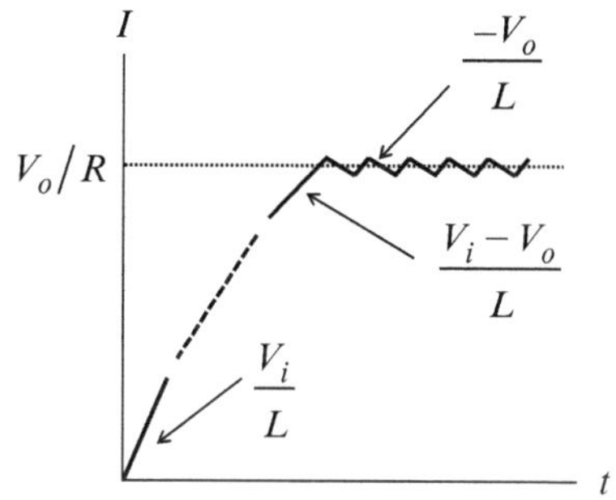

Fig. 3.13 Inductor current charged from the cold boot

oversampling comparator makes the 1b error polarity decision for the Δ function while the inductor L integrates the error voltage to perform the Σ function. As the current is switched, the capacitor C^* is to hold the output voltage during transients temporarily. Although it can also reduce the ripple further, too large loading capacitor makes the loop sluggish and unstable.

With the flux in the inductor empty, there is no current flowing at the beginning since it is a current storage device. The empty or low current state is defined as a turned-off or sleep mode in power supply uses. The inductor current should be charged before any load current can be drawn. That is, the input voltage V_i starts to be integrated, and the inductor current rises with a steep slope of V_i/L as shown in Fig. 3.13, which illustrates how the inductor current is ramped up.

During this cold boot period, the pulse-width or pulse-density modulated feedback pulse will exhibit almost 100 % duty cycle or mostly 1's sequence so that that the inductor current can be charged up. As the inductor current approaches the final average load current of V_o/R, the voltage on the inductor decreases, and the slope $(V_i - V_o)/L$ becomes less steep as the duty cycle or pulse density decreases accordingly. When the output exceeds the reference voltage, the duty cycle or pulse density decreases further, and the oversampling feedback stops charging and reverses the course by discharging the inductor current. That is, the inductor input stays switched to V_i longer on average than to ground when charging, but switched to ground longer than to V_i when discharging. These charging/discharging cycles make a controlled oscillation called ripple around the average DC load current. If the inductor input is switched to ground, and the current is discharged with a negative slope of $-V_o/L$.

The ripple is just formed for the feedback loop to sustain a constant DC average current delivered to the load. When the load varies, the output voltage rises or drops faster, and the duty cycle or pulse density decreases or increases accordingly. As a result, the inductor average DC current settles to a new level incrementally to counter the effect. Once the inductor is fully charged up to the normal current level, the feedback based on the oversampling principle monitors the output error, and generates either pulse-width or pulse-density modulated error. This pulse-shaped digital feedback signal changes the current by modulating either the duty cycle or the pulse density, which represents the duration of the pulse with a fractional time unit relative to 1.

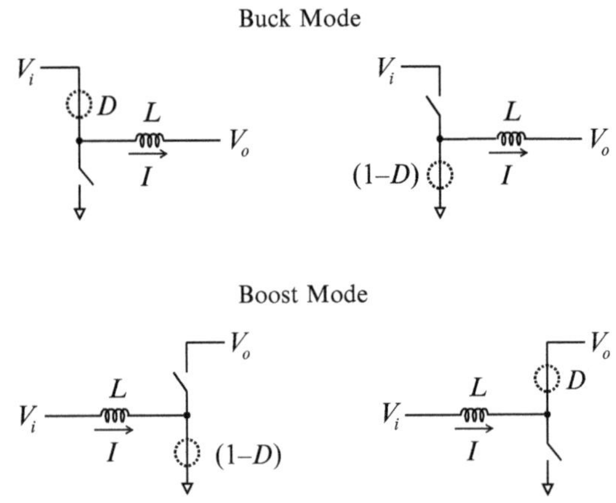

Fig. 3.14 Inductor current charged from the cold boot

As shown, the current of the inductor is charged up or down depending on the terminal voltage switched across it. If the voltage at the inductor input is switched to be higher or lower while keeping the output side voltage constant, the effective input voltage is smaller than the voltage switched since the duty cycle is smaller than unity. It is the DC–DC down-conversion (buck). On the other hand, if the input side is fixed, and the output side is switched, the divided output yields a higher effective output voltage, which is the DC–DC up-conversion (boost). Two inductor switching sequences are shown in Fig. 3.14 for the buck and boost modes, respectively.

The dotted circles mark the switches with their duty cycles. That is, the average output voltage and its ripple are directly related to the pulse duty cycle since the energy transfers into and out of the inductor should be balanced in the steady state.

In the buck converter, the oversampling modulator compares the output voltage with the reference, and controls the switch to the input with a duty cycle of D. The inductor current will be charged and discharged meeting the following conditions.

$$\begin{aligned}&\text{When switched to input}: V_\mathrm{i} - V_\mathrm{o} = L\frac{dI}{dt} > 0.\\&\text{When switched to ground}: 0 - V_\mathrm{o} = L\frac{dI}{dt} < 0.\end{aligned} \tag{3.11}$$

The inductor current is charged up and down as the inductor is switched to the input and to the ground. Thus the following relation can be obtained by equating the currents for charging and discharging.

$$\frac{V_\mathrm{i} - V_\mathrm{o}}{L} \times D = \frac{V_\mathrm{o}}{L} \times (1 - D), \tag{3.12}$$

which gives the input–output relation of

$$V_o = DV_i. \tag{3.13}$$

Since $D < 1$, it performs the buck conversion. If there is no power consumption in the inductor and the switch, there is no power loss in this DC–DC conversion. Thus the output power should be the same as the input power.

$$V_i I_i = V_o I_o + \text{Inductor/Switch Loss} \approx V_o I_o. \tag{3.14}$$

From (3.13) and (3.14), the following relation can be derived.

$$\frac{V_o}{V_i} = \frac{I_i}{I_o} = D. \tag{3.15}$$

This implies that the DC voltage is down-converted by the ratio of duty with no power loss. Note that (3.15) is the relation of the DC voltages and currents. The transformer works in the same way for the AC voltages and currents depending on turn ratio.

The boost converter works with the same principle. The inductor is charged with the input voltage, and discharged to the load with a duty cycle of D. Then,

$$\frac{V_1 - V_O}{L} \times D = -\frac{V_1}{L} \times (1 - D). \tag{3.16}$$

$$V_i = DV_o. \tag{3.17}$$

Since $D < 1$, it performs the boost conversion. So the following relation holds true, and the output can be higher than the input.

$$\frac{V_o}{V_i} = \frac{I_i}{I_o} = \frac{1}{D}. \tag{3.18}$$

The power efficiency of the switched-inductor buck or boost DC–DC converter can reach 100 % ideally, but considering the loss in the inductor and switch, it is limited to

$$\eta = \frac{R}{R + R_L + R_{SW}}, \tag{3.19}$$

where R_L and R_{SW} are the resistances of the inductor and the switch.

In practice, the power efficiency gets lower since additional power is consumed for all glue functions such as V_{ref} generator, error comparator, and feedback circuit for error control. For example, if a temporary storage capacitor C^* such as in Fig. 3.12 is used, the energy to charge it will be lost though the voltage ripple at the output is small. Also the continuous-time LDO with a large storage capacitor is required for all voltage regulators. Although the efficiency of the switched-inductor DC–DC converter is high, the overall power efficiency tends to go lower.

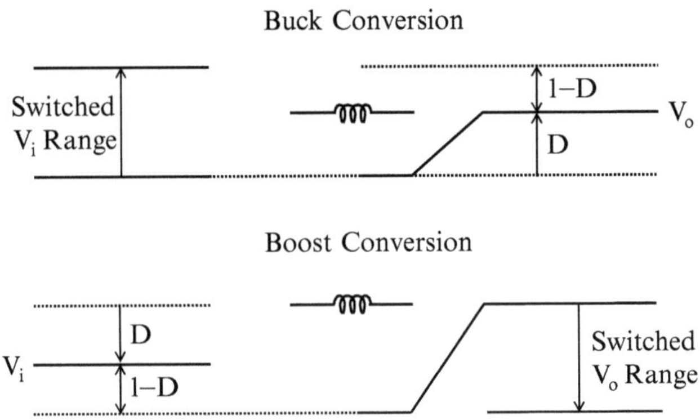

Fig. 3.15 Cold starting the buck-and-boost modes

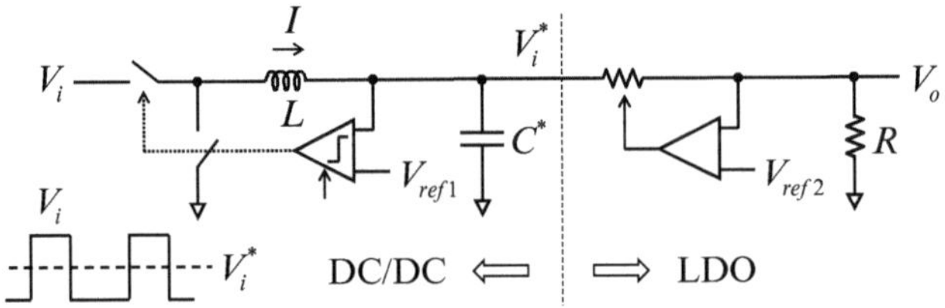

Fig. 3.16 Switched-inductor buck converter with LDO

Therefore, for low-power uses, even low-efficiency switched-capacitor DC–DC converters matter little.

Both buck and boost converters need cold starting as shown in Fig. 3.15. Usually, the cold start of the buck converter is straightforward. From the initial condition of zero current, the inductor can be charged with a full duty until the current reaches the desired level. Once it reaches the final value, the feedback based on pulse-width or pulse-density error modulation forces the inductor current to track the constant value. On the other hand, the initial cold start of the boost converter is a little tricky, and needs close attention. The inductor current and output voltage are valid at different clock phases. The inductor current can be charged up to the nominal level, but the output voltage stays low unless the boost current is switched to the output. If the output voltage is low, the negative feedback error keeps on demanding that the inductor current be charged by setting $D = 1$. That is, the inductor current is charged, but the output voltage is stuck at low voltages. Both in pulse-width and pulse-density modulations, the low-duty boost cycle is required, and the feedback control shouldn't allow the long sequence of $D = 1$. Once the output voltage exceeds the desired level, the feedback loop can take over.

Figure 3.16 shows an example of the switched-inductor buck converter with an LDO stage. The intermediate output V^* of the DC–DC converter is close to the output V_o to increase the power efficiency. Two V_{ref}'s set the levels of V_i^* and V_o, respectively. The buck converter averages out the pulsed input V_i with a duty cycle to be V_i^*. Including the LDO, the power efficiency given by (3.19) can be modified to be

$$\eta = \frac{R}{R + R_L + R_{SW}} \times \frac{V_o}{V_i^*}. \tag{3.20}$$

That is, there is no energy loss when charging the inductor flux with voltage except for the loss due to the inductor and switch resistances. The boost operation has the same high-power efficiency since the intermediate voltage is still higher than the output.

3.6 Glitch in DC–DC Converters

Modern portable or automotive electronic gadgets need multiple power supplies with very high-power efficiencies, and switched-inductor DC–DC converters are commonly used for that. Although the switched-capacitor DC–DC converter is handy to perform low-power on-chip signal processing such as DC multiplication or division, it is lossy and inappropriate to use for high-power supplies. Riding the miniaturization trend, even the power supplies are critically designed with shrinking dimensions, and the glitches in the power lines have been to blame in most electronics' failures. In fact, properly designed semiconductor chips rarely fail if the PVT (process, voltage, and temperature) conditions are met. Power failures are not rare in power-management chips since power in modern electronics is never turned off, but stays in the sleep mode waiting to wake up.

Assume the supply is in the sleep mode. If capacitors are used for energy storage, the sudden surge of the load current makes a voltage drop equivalent to $\Delta Q/C$ as explained in Fig. 3.3. As long as the capacitor is large enough to hold enough charge, the ripple can be made small. That is, the voltage ripple due to the transient output load can be abated effectively using only capacitors. By duality, the current ripple can be reduced using inductors. Therefore, if inductor is used for energy storage, the supply voltage glitch is more likely to occur as explained in Fig. 3.17.

The sleep mode is the state that the inductor holds a minimum current delivered to a reduced load to save power, but the voltage still remains high. If the load draws a large amount of currents like a step function for the normal operational mode as marked by a dotted line, the current in the inductor will be quickly drained out causing the voltage to drop together if $\Delta\psi/L$ exceeds the standby current level. To avoid the glitch situation, the inductor-based power supply should ramp up the storage current slowly as marked by a dashed line. The maximum current slew rate when the inductor is charged is V/L. The same glitch condition can be reached if the source battery doesn't hold enough charges as in automobile batteries. When large

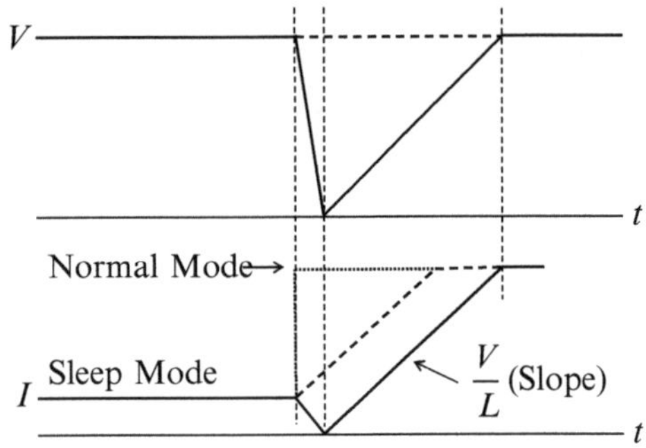

Fig. 3.17 Power supply glitch in switched-inductor DC–DC converters

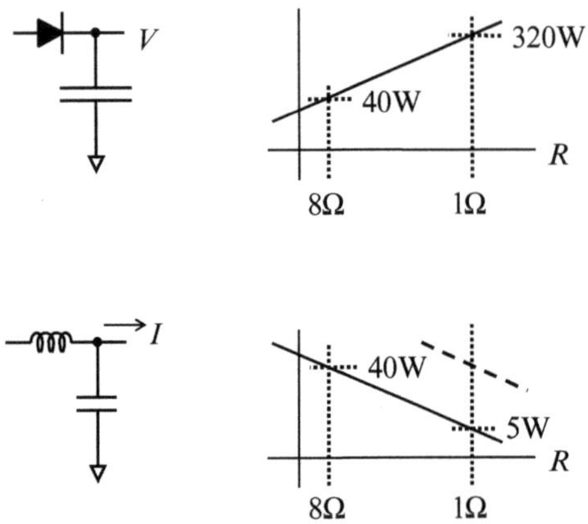

Fig. 3.18 Constant-voltage vs. constant-current power supply

amounts of current are drained by starting various functions in the automobile electronics, marginal poorly charged 12 V car batteries can suddenly dip down to as low as 7–9 V. This dip causes the current glitch, and can easily reset various power supplies in the automobile system.

Figure 3.18 shows examples of two power supplies. Audio power amplifiers are power supplies. One is a constant-voltage power supply using a large capacitor while the other is a constant-current power supply using a small inductor. They deliver high currents to low-impedance loads. They usually operate in the class AB push–pull mode to save power. If the audio signal is fed to the reference voltage, the power supply output will follow the input while delivering high power to the load. To reduce heat from the power dissipation, the switching power amplifier is commonly used in the form of the H-bridge, which is the differential form of the widely used pulse-width modulated audio signal. If they are used to drive an 8 Ω speaker whose impedance can dip as low as 1 Ω, the constant-voltage source amplifier can source far more power for low-impedance loads while the constant-current source amplifier fails unless large inductors supply high current as shown using a dashed line.

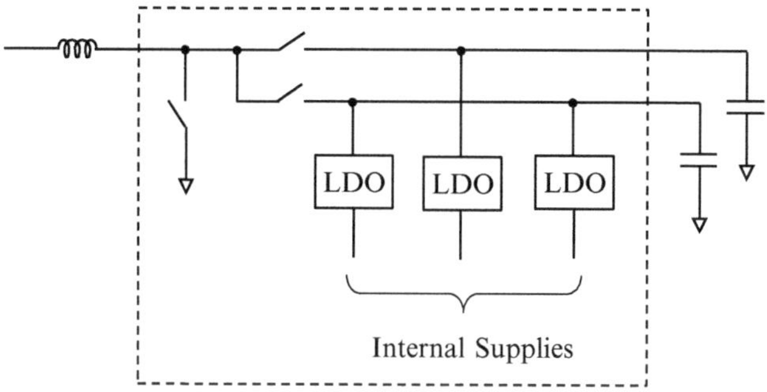

Fig. 3.19 Single-inductor boost converter with multiple on-chip LDOs

As given by (3.15) and (3.18), the output voltage of the switching DC–DC converter is set by the duty cycle of the inductor charging. Therefore, if the sum of the duty cycles is still less than 1, a single inductor can supply multiple DC voltages as shown in Fig. 3.19 [10, 11].

The ripple of each output can be filtered by a separate LDO. If the loading of the LDO is static DC, even the external capacitor of the LDO can be left out for the low bill of materials. However, without C_{ext}, the supply currents of the on-chip internal LDOs can be mixed at the inductor since the inductor memorizes previous currents. The retention circuits for continuous memory, control, or sensor shouldn't be tied to the DC–DC converter output directly as they should be operated all the times.

3.7 Almost DC Circuits for Body Sensors

Other than power supplies, circuits with almost DC spectrum are mostly for test instruments. Low-frequency testing instruments like voltmeter have almost capacitive inputs for high input impedance. Signals from human body such as ECG and EKG operate at low DC spectrum of Hz unit. Furthermore handling almost DC signals require an unusually high degree of DC accuracy. ECG signal itself is only mV level riding on 100's mV of CM signal. Its processing starts with a band-pass filter made of a high-pass and low-pass combination with cut-off frequencies of 0.5 mHz and 100 Hz, respectively. As it is severely influenced by the DC wander and motion artifact, very high dynamic range ADCs with over 16b ADC are used to accommodate an extra 6–7b (40 dB) DR for CM signals. Recently, most researches focus on the low-power instrumentation amplifier and SAR ADC. For low power, MOS transistors biased in subthreshold have been used still with analog anti-aliasing filters to reduce ADC sampling rates.

The old analog ECG sensor system is sketched in Fig. 3.20. The current trend is to integrate the whole system on chip. There are many sources that contribute to DC at the high-impedance sensor input. The body ECG signal is just one of them.

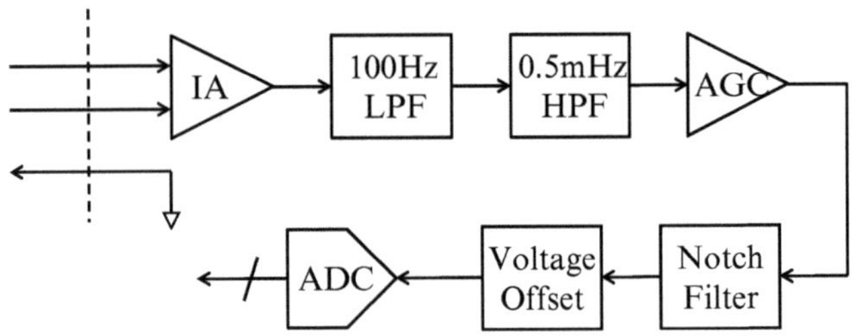

Fig. 3.20 ECG body sensor example

Fig. 3.21 Common-mode feedback for three-point sensors

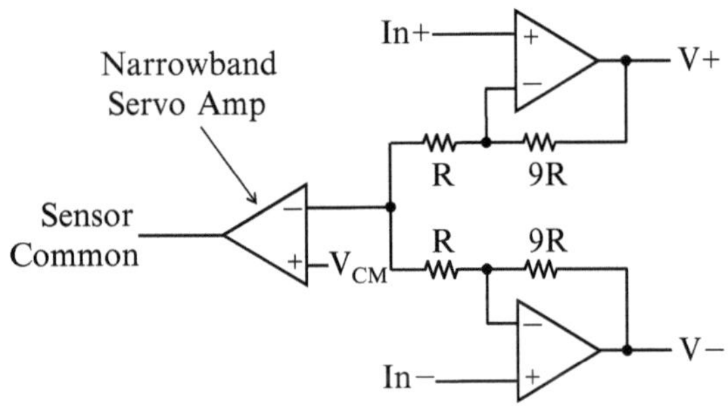

Fig. 3.22 Common-mode feedback using capacitive feedback

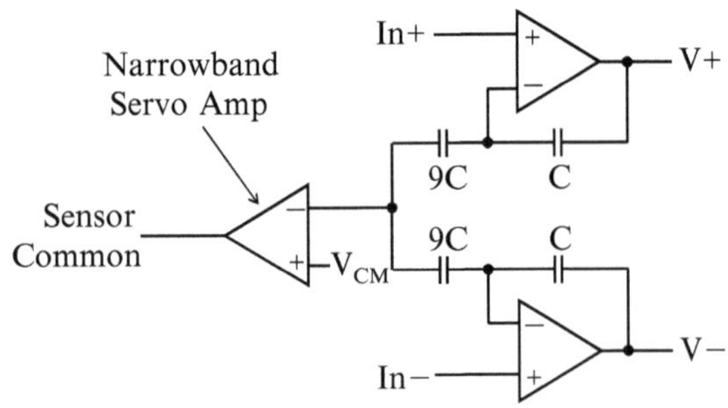

Others are body motion, sensor contact, 60 Hz, etc. If almost DC signal is high-pass filtered to eliminate its DC and CM, the extreme low-end of the signal is also lost, which causes the DC wander. That is, the DC operating point is undefined. The common implementation of the input instrumentation amplifier is shown in Fig. 3.21.

In the old three-point sensing system, the high-pass filtered input differential signal of In+ and In− is applied to two high-impedance opamp inputs. It is a difference circuit with a non-inverting gain. Its CM signal is sensed and compared to V_{CM}, and low-impedance narrowband servo amplifier forces the body CM voltage, which is usually tied to the ankle using a long wire. This instrument amplifier can be easily implemented in CMOS using capacitive feedback as shown in Fig. 3.22 [12].

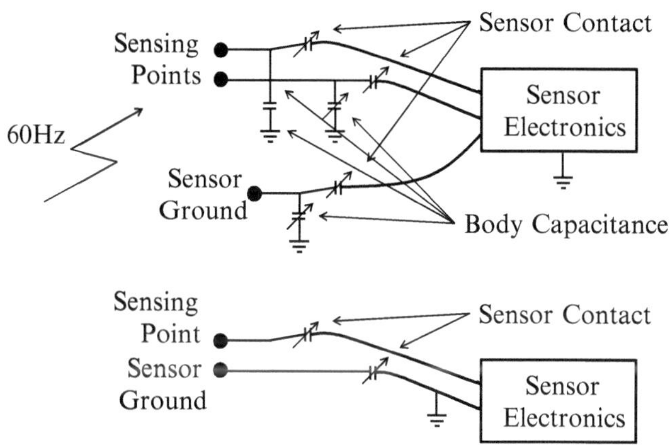

Fig. 3.23 Three-point vs. floating two-point body sensing

If everything else is assumed to be stationary except for the body ECG signal, it becomes the stationary channel, and there is no DC wander problem. However, the input DC of the dynamic channel is undefined. In the above case of capacitive input coupling, if the coupling capacitance changes due to the body motion, the sensor input DC changes suddenly. Like the old DC voltmeter, a need to reset the high-impedance input nodes arises. Furthermore, the capacitive instrumentation amplifier also suffers from the DC wander in the CM feedback loop. The DC path to the capacitive input node should be supplied. This high-gain CM feedback loop can lower the CM voltage swing, but the CM rejection ratio, which is defined as the CM signal gain to the differential output, is still limited by the matching accuracy of capacitors. That is, a volt-level CM swing can come out as a mV-level differential output.

Figure 3.23 illustrates the three-point vs. two-point sensing schemes. There are three capacitors at the sensing nodes. They are the body capacitance C_b, the sensor capacitance C_s, and sensor input capacitance C_i. If charge is conserved with no loss, the total charge on three capacitances is defined as

$$Q = C_b V_b + C_s V_s + C_i V_i, \tag{3.21}$$

where V_b, V_s, and V_i are voltages on the three capacitances, respectively. Now if any capacitances of C_b and C_s change due to body motion, some of their charges should flow into C_i to conserve the total charge Q, which can be defined as the effect of the motion artifact, which results because the voltage on the high-impedance capacitive node is undefined. There can be so many capacitive coupling paths, and such changes would make the sudden change of the DC at the sensor input. If long wires are used, even 60 Hz power line is picked up by the sensing node.

The floating two-point sensing shown at the bottom can greatly reduce this effect. If the probe has a direct ohmic contact, the two-point floating ground system has no DC problem. However, if it accommodates both resistive and capacitive sensor input probes, the same high-impedance node problem arises at the sensor input though at a lower level. In the floating sensor, the DC uncertainty due to the capacitive coupling through C_b doesn't exist. If the input changes too much by the capacitive coupling, which is possible when the coupling capacitor is too small, the sensed signal has no sufficient ECG information. It is as if the high-impedance at the input is open. The high-frequency energy can be also captured by long wires working like antenna, and the sensor dynamic range should be extended to cover that. However, most low-frequency coupling can be capacitive, and short shielded wire will eliminate the antenna effect. Typically, the sensor dynamic range with extra 20 dB headroom can cover this capacitive DC motion artifact. If the coupling energy exceeds the head room, the sensed signal may not be meaningful at all.

If the resistive probe is used, there is no DC at the input since the ground of the sensor directly connected to the body, and the relative voltage of the other probe point can be sensed. However, for capacitive sensing, the input needs to be reset to define the DC input of the circuit. Note that the only viable way to reduce the DC effect is using the correlated double sampling (CDS) as discussed in Chap. 1. The sensor input should be reset to a fixed DC periodically. In imager applications, they compared the input voltage to a certain DC level when the signal is at the black level, and use an up/down charge-pump to keep the input voltage constant. It is the same situation that the baseline of the ECG signal can be set to the constant level like the black level in the imager.

The image signal has an undefined DC fluctuation dominantly due to the 1/*f* noise of the sensor. The 1/*f* noise of the ECG sensor is also critical though it can be removed by analog offset cancellation or by chopping the front-end AGC amplifier. The AC coupling by high-pass filtering is not the solution. In the 3-point conventional sensor with 18–20b ADCs, the most common way is to chop amplifier offsets. However, in the two-point sensor, digital cancellation using CDS would suffice. The CDS is the relative signaling scheme for DC wandering systems like imager. Assuming that the black DC level is constant, the line signal is referenced to this level rather than the absolute level. So it collectively removes the slowly-varying almost DC components in the input and the 1/*f* noise of the circuit.

Two-point sensor is shown conceptually in Fig. 3.24. The scheme of floating ground eliminates the CM signal. Analog anti-aliasing filters can be replaced by digital filters. The periodic reset and digital CDS eliminates DC wandering.

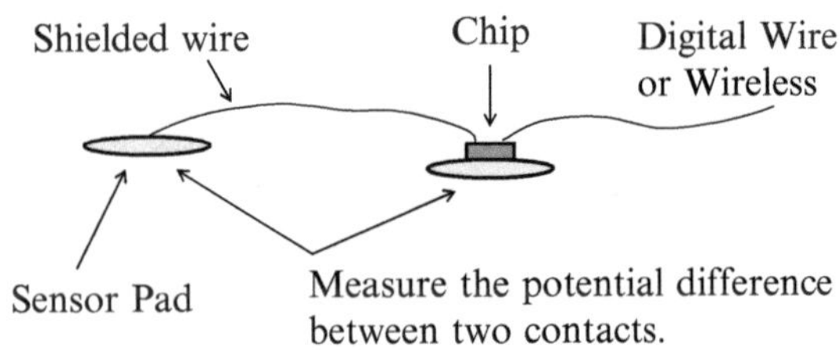

Fig. 3.24 Two-point sensor implementation

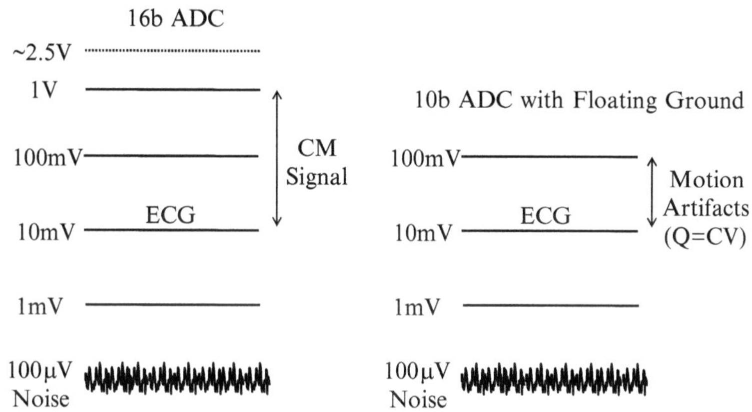

Fig. 3.25 ADC dynamic range for ECG signal and noise

The mV-level signal can be amplified up to ~10 mV level using a continuous-time ΔΣ modulator with a 100 mV input range. Two design constraints are the global AGC and 20 dB headroom for motion artifact. For AGC, three levels of peak, bottom, and baseline should be detected to fit the signal into the ADC range. It is true that other artifacts can result from the periodic reset, but some digital ways can be figured out to smooth out the reset transients, or CDS can be undone digitally. Since the input is periodically reset, the constant reset error can be used to recover the original smooth waveform if reset occurs between the peaks. Similarly, the baseline DC wander can be digitally corrected so that after each period the baseline can be made to stay the same. This implies that the ECG signal can be captured with high DC uncertainty and $1/f$ noise, but reshaped digitally.

Figure 3.25 sketches the dynamic range budget of the ADC for the proper ECG signal acquisition. The ECG signal is at about mV level if the probe makes a good capacitive contact and the signal is assumed to be preconditioned using AGC. Within 100 Hz bandwidth, the noise level of the sensor circuit can be as high as 100 μV level dominated by $1/f$ noise. In old three-point sensing on the left side, almost 40 dB extra dynamic range is reserved for high CM voltages, and thus an ADC with over 16b dynamic range has been used. For further digital signal processing such as digital AGC and amplification, more than 18–20b dynamic range is required. However, in floating two-point sensing, the dynamic range requirement is significantly relieved as there is no CM swing. It can be fitted into the 10b ADC dynamic range as shown on the right side. To achieve this low dynamic range, the SAR ADC can be used, but it suffers greatly from high wideband sampled noise and latch hysteresis, thereby requiring an extra 20 dB gain to get a volt-level signal. It also requires a steep anti-aliasing filter. For this purpose, the continuous-time ΔΣ modulator is an ideal choice.

First, the oversampling continuous-time ΔΣ modulator makes it possible to move the anti-aliasing function into the digital domain, and its low noise performance enables direct quantization at 1–10 mV level with 12b dynamic range

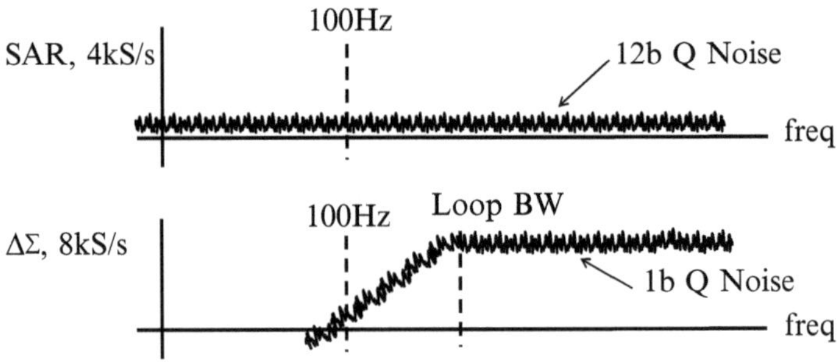

Fig. 3.26 In-band quantization noises of SAR vs. ΔΣ modulator

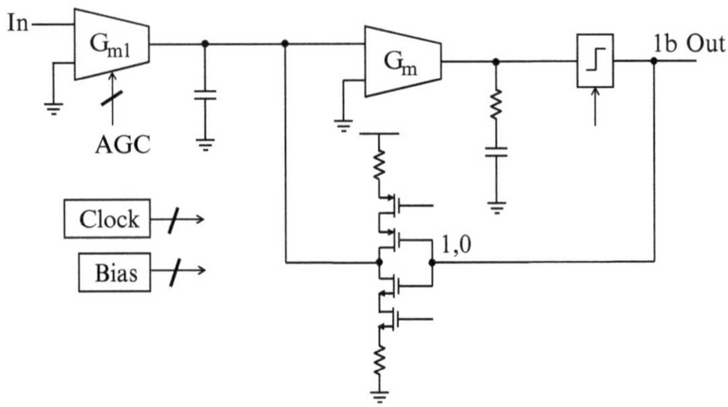

Fig. 3.27 Second-order continuous-time ΔΣ modulator

without amplifying the input further to a volt level. Also it is much easier to achieve 12b linearity without requiring 12b component matching. Furthermore, designing 1b ADC greatly simplifies the comparator design. The quantization noises and sampling rates for SAR and second-order ΔΣ modulator are beyond comparison as shown in Fig. 3.26.

A second-order ΔΣ modulator is sketched conceptually using G_m cells as integrator in Fig. 3.27. It is a minimal architecture made of two G_m cells, a comparator, and a charge-pump. The charge-pump performs as a feedback DAC and saves the bias current of a class-A G_m cell. The series resistance with the integrating capacitor is to add a zero for loop stability. Even by programming the input G_m cell, the AGC function can be integrated into it. Since the comparator is a 1b ADC, it doesn't require any preamplifier. That is, this modulator replaces the whole chain of the ECG sensor circuits shown in Fig. 3.20. It performs the functions of the instrumentation amplifier, and eliminates all remaining functional blocks for 0.5 mHz HPF, 100 Hz LPF, AGC, 60 Hz notch, offset, and ADC without requiring any 12b matching and high preamplifier gain.

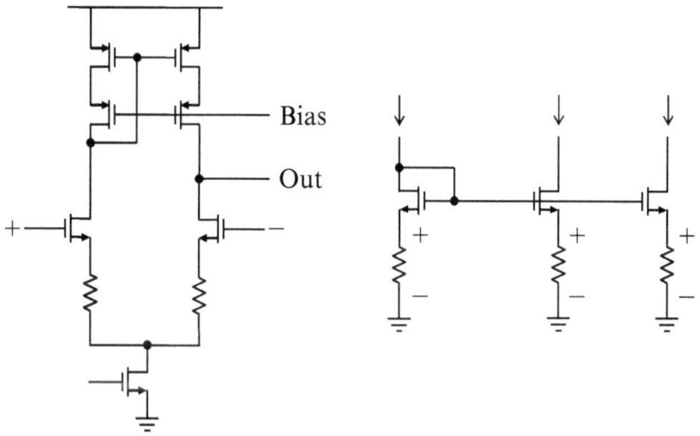

Fig. 3.28 Gm-C integrator and current mirror

The low-power G_m cell can be made of a simple differential pair with source degeneration as shown in Fig. 3.28. With the low bias current level approaching the subthreshold region, the matching property of MOS devices degrades rapidly. In particular, MOS transistors in current mirrors do not match well. The source degeneration makes the G_m cell far less insensitive to process, and the voltage drop across resistor improves the current source matching.

References

1. K. Leung, P. Mok, A capacitor-free CMOS low-drop-out regulator with damping-factor-control frequency compensation. IEEE J. Solid State Circuits **38**, 1691–1702 (2003)
2. M. El-Nozahi, A. Amer, J. Torres, E. Sanchez-Sinencio, High PSR low drop-out regulator with feed-forward ripple cancellation technique. IEEE J. Solid State Circuits **45**, 565–577 (2010)
3. APP725, DC-DC conversion without inductors, *Maxim Integrated* (July 2009)
4. W. Kester, B. Erisman, G. Thandi, Switched-capacitor voltage converters, in *Analog Devices*, Section 4, Handbooks (1998)
5. H. Le, M. Seeman, S. Sanders, V. Sathe, S. Naffziger, E. Alon, A 32nm fully integrated reconfigurable switched-capacitor DC-DC converter delivering 0.55W/mm2 at 81% efficiency, *ISSCC Dig. Tech. Papers* (Feb 2010), pp. 210–211
6. V. Ng, M. Seeman, S. Sanders, High-efficiency, 12V-to-1.5V DC-DC converter realized with switched-capacitor architecture, in *IEEE Symposium on VLSI Circuits*, June 2009, pp. 168–169
7. L. Svensson, J. Koller, Driving a capacitor load without dissipating fCV^2, in *IEEE Symposium on Low Power Electronics*, Oct 1994, pp. 100–101
8. B. Ginsburg, A. Chandrakasan, An energy-efficient charge recycling approach for SAR converter with capacitive DAC, in *IEEE International Symposium on Circuits and Systems*, May 2005, pp. 184–187
9. M. van Elzakker, E. van Tuijl, P. Geraedts, D. Schinkel, E. Klumperink, B. Nauta, A 10-bit charge-redistribution ADC consuming 1.9μW at 1MS/s. IEEE J. Solid State Circuits **45**, 1007–1015 (2010)

10. H. Le, C. Chae, K. Lee, S. Wang, G. Cho, A single-inductor switching DC-DC converter with five output and ordered power-distributive control. IEEE J. Solid State Circuits **42**, 2706–2714 (2007)
11. M. Huang, K. Chen, Single-inductor multi-output (SIMO) DC-DC converters with high light-load efficiency and minimized cross-regulation for portable devices. IEEE J. Solid State Circuits **44**, 1099–1111 (2009)
12. L. Turicchia et al., Ultralow-power electronics for cardiac monitoring. IEEE Trans. Circuits Syst I **57**, 2279–2289 (2010)

Chapter 4
Data-Converter Circuits

Data converters perform two functions: Data acquisition and data distribution. The former is to acquire digital data from analog channels for digital signal processing, and the latter is to distribute the processed digital data back to the analog channels. Two essential elements for these tasks are analog-to-digital converter (ADC) and digital-to-analog converter (DAC). There are various types of converters with a wide range of sampling rates and resolutions. Data converters help to overcome the flaws of switched-capacitor analog sampled-data processing, and make digital signal processing possible. Since it is not affected by analog imperfections, digital signal processing has replaced analog signal processing. Accuracy of the data converter is limited by magnitude and transient errors in the sampling and signal generation. The magnitude error is a static DC error that can be trimmed or calibrated while the transient error is a dynamic error that needs only to be controlled precisely.

4.1 Data Acquisition and Distribution

Sampled-data analog signal processing is an analog equivalence to the discrete-time digital signal processing. Without exceptions, both ADC and DAC require prior and post filters called anti-aliasing filter (AAF) and smoothing filter (SF), respectively. They limit signal to be contained inside the Nyquist band. In addition, all wide dynamic range (DR) data acquisitions require an automatic gain control (AGC) function at front as most ADCs have a limited spurious-free dynamic range (SFDR) as shown in Fig. 4.1.

The data acquisition performs the most common electronic function of quantization used from simple sensors to wireline or wireless receivers while the data distribution is to reconstruct signal. They exhibit a wide range of performance, and are often characterized by their sampling rate, power, resolution, chip area, noise, monotonicity, SFDR, etc. There are two types of them. One is sampling at the

B.-S. Song, *System-level Techniques for Analog Performance Enhancement*,
DOI 10.1007/978-3-319-27921-3_4

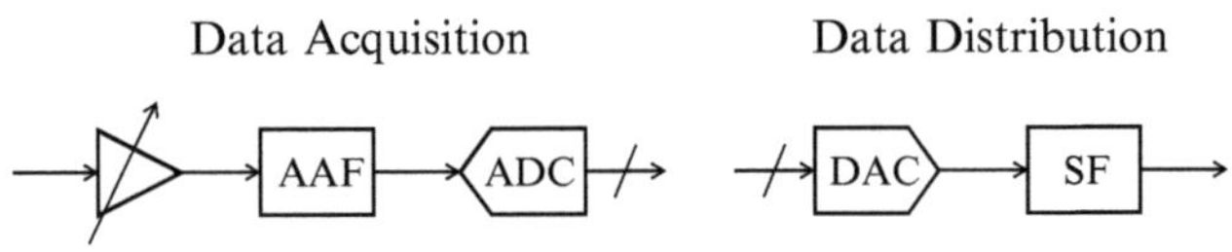

Fig. 4.1 Data-acquisition and data-distribution concepts

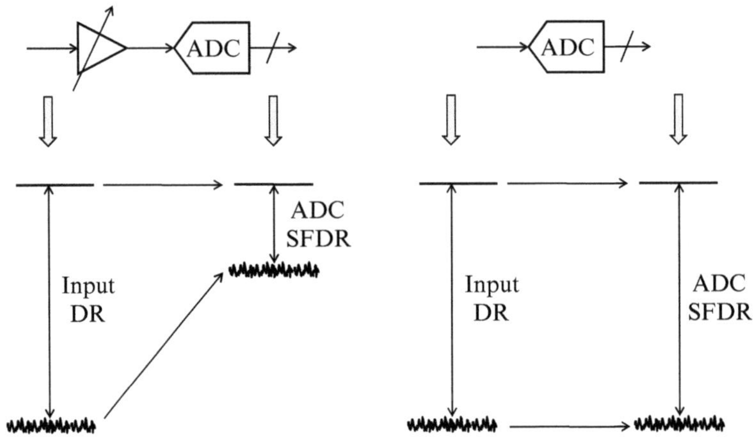

Fig. 4.2 AGC extends ADC dynamic range

Nyquist rate, and the other is oversampling. The former is an open-loop system that requires absolutely accurate offset, gain, and component matching while the latter is a closed-loop feedback system that trades loop gain for performance.

The performance of data-acquisition systems heavily depends on how to balance the dynamic ranges of AAF, AGC, and ADC. For example, modern high-resolution image sensors don't require the AAF since the pixel signal is already sampled. However, they still require a wider than 14b dynamic range to resolve from very bright to dark images, and in particular, to tell any dark portion of the image from the darker part. This can be implemented with either a 30 dB (5b) AGC and a 9b ADC or with a 14b ADC doing AGC digitally as shown in Fig. 4.2.

In old designs, there was no choice other than using analog AGC and low-resolution ADC. However, recent advancements of CMOS analog techniques along with fast digital switching enable designers to configure such data-acquisition systems in the digital domain. The most notable recent trend is towards implementing the high-resolution, high-speed ADCs with high SFDR, thereby performing digital AAF, channel filtering, AGC, and decision all digitally.

Figure 4.3 shows three choices of data acquisition or receiver systems, where the desired channel is in the middle of the spectrum with nearby interferences. In RF receivers, they are also called blockers. In the top case, to precondition the signal for low-resolution flash ADC, it requires significant analog signal processing that includes AGC, AAF, and even partial channel filtering to obtain the blocker-free

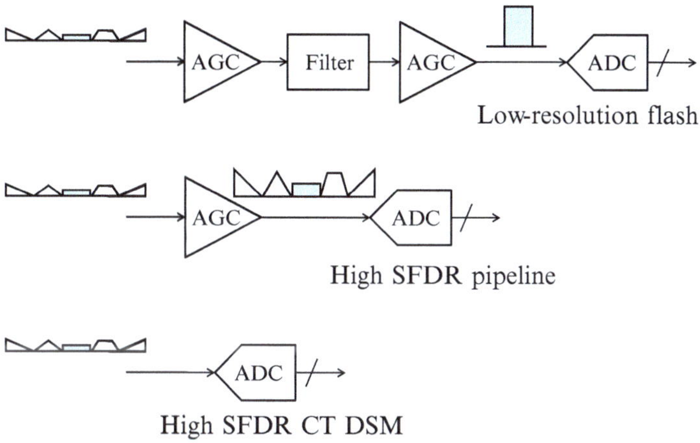

Fig. 4.3 System configurations for data acquisition

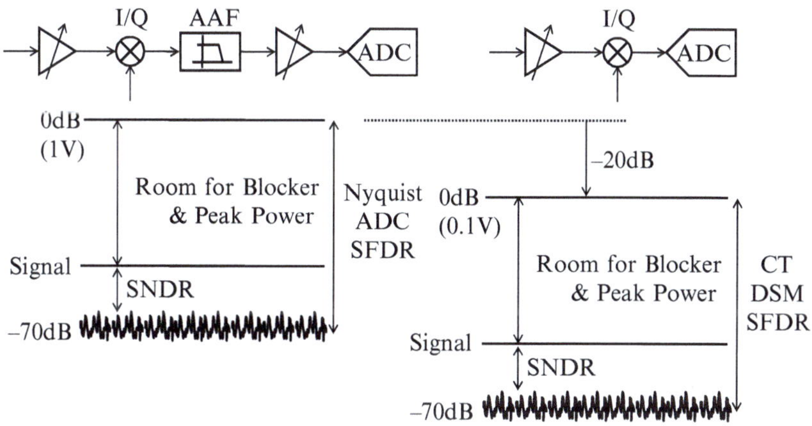

Fig. 4.4 Nyquist ADC vs. CT DSM SFDR

signal. Some analog processing of filtering can be saved using high SFDR Nyquist-rate ADC such as pipeline as in the middle case. Nyquist-rate ADCs have typically higher noise levels than oversampling ΔΣ modulators (DSM). That is, the low-level signal can be quantized directly with DSM as shown at the bottom. The difference of about 20 dB makes a big impact in the RF receiver design as shown in Fig. 4.4.

The bottleneck of RF systems is the SFDR of the down-conversion mixer which typically is about 70 dB for 10–20 MHz bandwidth systems. RF signals suffer from adjacent or alternate-channel blockers typically 40–50 dB stronger. That is, digital data can barely pass through the in-phase and quadrature (*I*/*Q*) mixers. As shown on the left side, Nyquist-rate ADC still needs AAF and AGC to obtain a volt-level signal at the ADC input. On the other hand, continuous-time (CT) DSM can quantize the mixer output directly, saving further the amount of analog processing.

It is shown that the intelligent system partition and the choice of AGC and ADC schemes greatly affect the system performance in the software-definable radio environment. The CT DSM exhibits about a −20 dB lower noise floor, and requires no AA filtering.

4.2 Nyquist-Rate vs. Oversampling ADC

Over the years, fine lithography has offered the speed and matching advantage in the CMOS data-converter design, but with low voltage and gain constraints. Furthermore, low-power requirement has made designers search for alternative opamp-free methods relying more on digital switching such as successive approximation register (SAR), time-domain (TD), and comparator-based ADCs. They are open-loop Nyquist-rate ADCs that require absolute accuracy though digital correction can be applied to relieve the sub-ADC resolution requirement in limited uses. Digital calibration is to further enhance their static linearity performance.

4.2.1 Opamp-Based ADC

ADCs are comparator arrays making decisions on whether the sampled input is higher or lower than the finite number of equally spaced reference levels. The number of levels is the resolution represented by the binary number of bits N. The basic flash ADC makes all decisions in one flash shot. For high resolution, N gets larger, and the total number of comparators grows exponentially by $(2^N - 1)$. Therefore, the flash ADC is not desirable if both chip area and power consumption are considered, and seldom used for higher than 7–8 bits unless very high sampling rates are wanted. Most ADCs make decisions in multiple steps using the subranging concept. The residue is the unquantized portion of the signal left over after a finite number of bits are decided in one step. Data-converter circuits are about how to generate the subrange residue. Although the subrange residue can be generated without using opamp, it is advantageous to amplify the residue to relieve the accuracy requirement of the subsequent sub-ADC stages. Three basic functions of the residue amplifier are to sample, subtract, and amplify. Note that one capacitive switched-capacitor MDAC residue amplifier performs all three functions.

Figure 4.5 shows the standard linear amplifier and two examples of opamp-free dynamic nonlinear amplifiers in the time-domain. The capacitive switched-capacitor residue amplifier at the top purely operates in the small-signal linear mode [1, 2]. The analog error decreases exponentially in an analog way with a time constant related inversely to the dominant pole frequency of the amplifier. It is very linear and DC stable due to the negative feedback. The speed of settling depends on the frequency domain steady-state small-signal bandwidth. It is a DC circuit. Once it settles with enough accuracy, the only two remaining DC error components in the

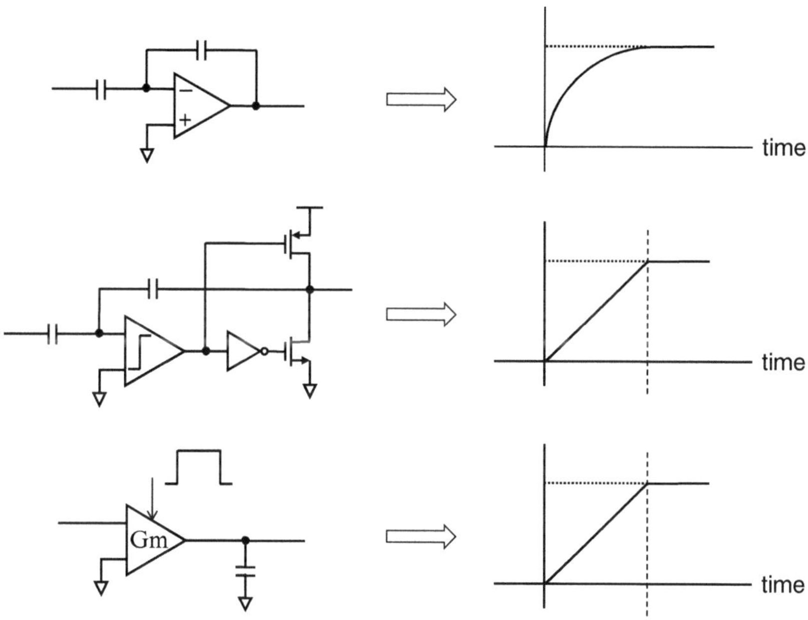

Fig. 4.5 Linear vs. nonlinear residue amplifiers

output result from the capacitor mismatch and the opamp gain and nonlinearity. For high resolution, the coarse ADC error can be corrected digitally using the overranging redundancy, but the capacitor ratio and opamp gain errors remain. They are fundamental, and the only way is either to trim or calibrate them in analog or digital ways.

On the contrary, the middle amplifier slews nonlinearly, and the output is frozen in time when the output stops slewing by detecting the zero-crossing [3]. Its accuracy entirely depends on the zero-crossing accuracy that requires an accurate self-latching operation. The self-latching amplifier to detect zero-crossing should have an infinite gain and wide bandwidth. The self-latching is an ideal comparator operating without clock. It is basically an analog track and hold circuit which is sensitive to nonlinear sampling and aperture delay errors. Such self-latching comparator has been used in the ADC version of the ripple-through pipelined, asynchronous SAR, or clockless ADC.

The open-loop integrator at the bottom is an analog version of the very popular digital integrate-dump function, which is a matched filter in digital communications to recover the impulse symbol spread over the symbol interval while averaging noise [4]. For impulse symbol, it offers a significant SNR advantage as signal is integrated but high-frequency noise is suppressed. However, if used for analog amplification, the SNR stays the same. In DC-unstable systems, SNR gets worse. The gain also depends on the output slew rate set by the active transconductance device and the integration time. Analog integrator is used only in the closed-loop circuit such as filter, but never used in the open-loop circuit due to the DC instability. The active device Gm is very nonlinear unless linearized by feedback, and its absolute value is inaccurate and undefined. The open-loop integrator is not a proper choice for precise residue amplification.

One trend with the dynamic amplifier is to switch the bias current off for low-power consumption [5]. It is just to power the amplifier only when it amplifies, and to turn off power when not in use. However, the bias circuit is a DC circuit with a very long time constant due to the decoupling capacitor. If it is turned on and off, it takes a very long time for the bias current to be fully turned on and stabilized. That is, the unsettled inaccurate bias affects the output settling accuracy, and the briefness of the bias current turned-on time does not allow enough time for the residue amplifier to settle.

4.2.2 Opamp-Free ADC

In the pursuit of high-resolution and high-speed ADCs, designers tumbled into the road block imposed by the high power consumed in the residue amplifier and also the difficulty of implementing accurate residue gain, which prompted designers to look for low-power alternatives from old ADC architectures such as SAR and slope-type ADCs. In the former, the residue voltage is generated using a charge redistribution DAC, and only the comparator makes decisions with a full N-bit resolution for $N+1$ cycles [6]. The latter is called time-to-digital converter ADC (TDC), and requires a linear voltage to frequency or phase converter. In systems which include the VCO as in PLL, TDC works advantageously to quantize the phase error directly. The common difficulty the opamp-free ADC faces is the stringent requirement for the quantization error as demanded in all open-loop Nyquist-rate ADCs. The other prominent development to greatly alleviate this requirement is the feedback approach that encircles the quantizer with a high-gain feedback loop and suppresses its quantization error by the loop gain. Since the feedback gain is possible with high oversampling rates, it is called oversampling $\Delta\Sigma$ modulator or noise-shaping coder [7–9].

The logic behind developing opamp-free ADCs is that low figure of merit (FOM) has now become an essential necessity, and it overshadows all other design issues [10, 11]. New technologies have always offered lower FOM at lower supply than previous generations. However, since high-performance ADCs require higher supply voltages, opamp-free architectures such as SAR, DAC-free TDC, and comparator-based ADCs should trade performance for low power and speed. Nonlinear transfer functions used in TDC such as voltage-to-current, voltage-to-frequency, and voltage-to-phase converters should be linearized by oversampling techniques [12].

4.2.3 Digital Correction and Feedback of Quantization Error

As discussed, the basic function of ADC is to make decisions with the quantization error reduced as low as possible. Since quantization error has a random white spectrum, it is often called quantization noise, but it is a deterministic error. It is getting more elusive to operate open-loop Nyquist-rate ADCs with a desired level of quantization error as sampling frequency goes higher. There have been only two systematic practical ways to reduce the quantization error. One is the digital correction scheme used in Nyquist-rate multi-step ADCs such as pipelined ADCs [13, 14], and the other is the feedback concept used in oversampling DSMs.

Figure 4.6 shows the quantization noise spectral densities of three cases. On the left side, the quantization error spectrum of the Nyquist-rate SAR is spread evenly over the signal band, and the in-band quantization error integrated over the Nyquist band should be smaller than $V_{\mathrm{ref}}/2^N$ when N-bit decisions are made. That is, all bit decisions are final. On the other hand, in the digitally corrected pipelined ADC in the middle, a redundant overrange of the residue can be quantized with a coarse sub-ADC with low resolution. Since the error bit occurring during the coarse decision is corrected, the sub-ADC can operate at a far lower resolution of only $V_{\mathrm{ref}}/2^N$ ($N = 1$–4). In the oversampling DSM on the right side, the quantization error spectrum is shaped by high-pass filtering so that only the integrated in-band noise can be lower than the required. Therefore, as in the pipelined ADC, the DSM can operate with a low resolution of $V_{\mathrm{ref}}/2^N$ ($N = 1$–4) but at an oversampling rate since the resolution is enhanced by the feedback loop gain. The amount of feedback loop gain depends on the oversampling ratio and the order of the modulator.

DSM is based on the high-gain active filter given by integrators, which requires no accurate gain. The integrator output is linearized by the feedback loop gain. However, to make use of the oversampling advantage, the oversampling clock should be set higher by the factor of 2π than the unity-gain frequency of the loop filter. Therefore, the main constraint of the DSM operation is the minimum oversampling ratio of π. The higher the sampling rate gets, the more the in-band quantization error is lowered. The higher order of the loop filter also makes the slope of the high-pass filter steeper, and reduces the in-band quantization error to a greater degree. In effect, one pole can add an extra +6 dB/octave slope to the noise transfer function (NTF). The remaining high out-of-band quantization noise spectrum in the DSM output can be digitally filtered.

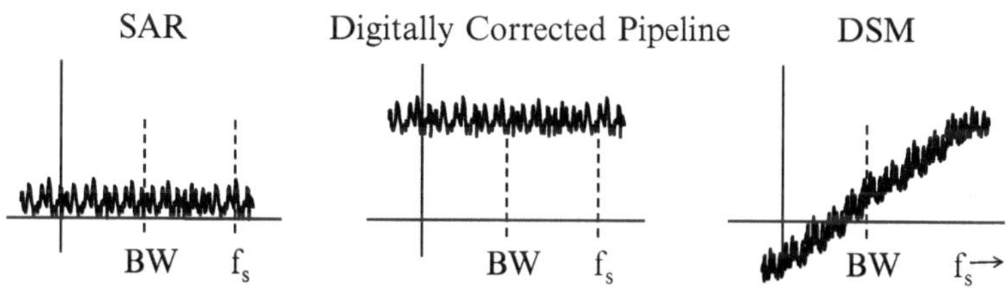

Fig. 4.6 Three quantization error spectral densities

4.2.4 Noise Implications in Data-Converter Circuits

In data-converter circuits, there are four types of noises. They are quantization, aliased wideband, thermal, and kT/C noises. As discussed, the quantization error is deterministic, but its spectral density is evenly spread like random white noise. Thus it is handled like a random noise. The wideband noise aliased during sampling originates from the previous stage, and ADC designers tend to ignore its existence. It is worse if the previous stage output is broadbanded. Whether circuit is switching or not, the thermal noise is fundamental, and affects the circuit performance continuously. Since the kT/C noise originates from sampling like the aliased wideband noise, it sets the lower limit of the noise level of sampled-data circuits.

Sampled-data circuits are made of two circuit elements, amplifier and integrator. Their operation starts from sampling signal on the capacitor. Therefore, the MOS switch with low on-resistance is needed to sample signal with full accuracy within a given sampling period. In the sample/hold, the noises from the finite switch on-resistance are marked as shown in Fig. 4.7.

During the sampling phase, switch noises are sampled on the capacitor band-limited only by the RC time constant, which yields the sampled kT/C noise. During the hold phase, the capacitor is flipped back and connected to the opamp output forming the unity-gain feedback. In amplifiers, the unity loop-gain frequency is lowered by the feedback factor. The switch noise during this hold phase is band-limited by the unity loop-gain-bandwidth f_{BW}, and doesn't contribute to the output noise since the RC time constant is typically much shorter than the feedback loop time constant of $1/f_{\mathrm{BW}}$. The next stage samples this output together with the opamp output thermal noise. Thus, the kT/C noise of the sample/hold stage is close to kT/C.

In the integrator case shown in Fig. 4.8, the switch noises appear in both clock phases. Therefore, the kT/C noise power is doubled, and the equivalent input-referred noise spectral density is the same as the thermal noise of the effective resistance of the switched capacitor, $R_{\mathrm{eff}} = 1/f_{\mathrm{c}}C$.

$$\frac{v_{\mathrm{i}}^2}{\Delta f} = \frac{2 \times \frac{kT}{C}}{f_{\mathrm{c}}/2} = \frac{4kT}{f_{\mathrm{c}}C} = 4kTR_{\mathrm{eff}}. \tag{4.1}$$

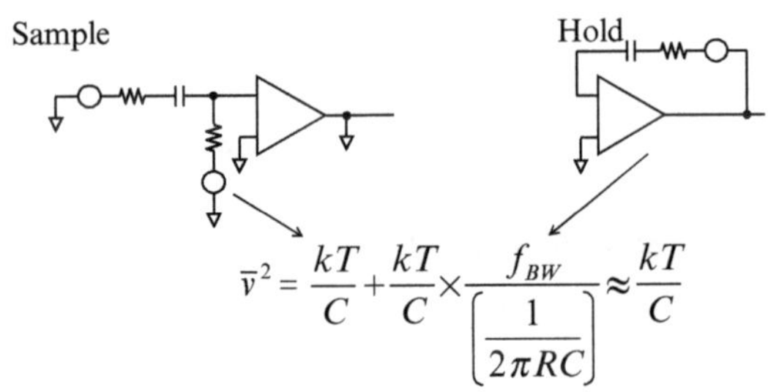

Fig. 4.7 Noise in sample/hold circuit

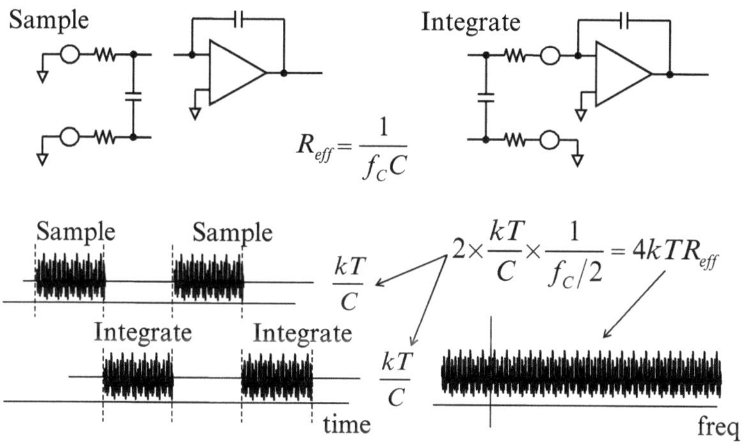

Fig. 4.8 Noise in integrator

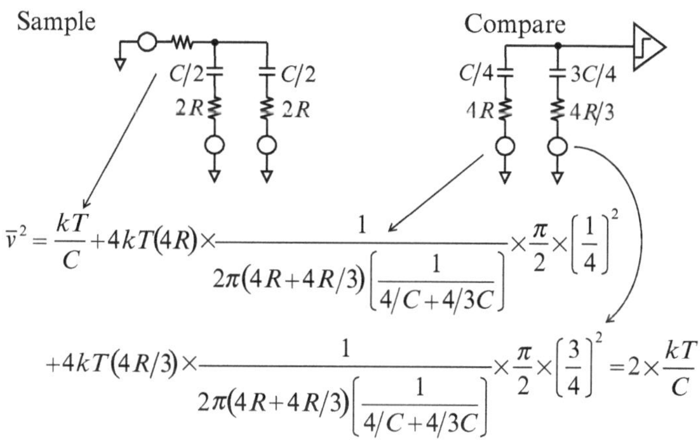

Fig. 4.9 Noise in SAR

However, the *kT*/*C* noise during the integration phase is band-limited by the integrator as in the sample/hold case, and the *kT*/*C* component sampled with the input is dominant.

The signal and the sampled *kT*/*C* noise is a fixed DC component at the integrator input while the switch noise is an AC spectrum. In the continuous-time integrator with an effective resistance R_{eff}, both signal and noise are AC. Therefore, for switched-capacitor integrators, (4.1) is true for the input-referred noise, but at the integrator output, it is closer to $2kTR_{eff}$ in effect. Similarly, as in the integrator case, the *kT*/*C* noise of the SAR appears in both sampling and comparison phases as shown in Fig. 4.9.

If the total capacitance of the SAR DAC is *C* and the *RC* time constants are scaled to be constant per each capacitor, the total *kT*/*C* noise power is doubled when referred to the comparator input as the sampled noise is added during the comparison phase. However, if the comparator preamplifier is band-limited, the *kT*/*C* noise component during the comparison phase becomes negligible as in the integrator case. The switch noise during the amplification, integration, or comparison phase is also equivalent to *kT*/*C* if the sampling bandwidth is only set by its own *RC* time constant. However, the signal bandwidth imposed by the feedback opamp is typically narrower than the bandwidth of the sampling network, and the *kT*/*C* noise is mainly set by the sampling capacitor size.

4.2.5 Nyquist-Rate SAR vs. Oversampling CT DSM

A perennial question remains as to Nyquist-rate vs. oversampling ADCs. Except for the sampled-data acquisition for image and touch-pad pixel data, most data-acquisition systems include signal preconditioning functions such as AAF and AGC. In modern wireless receivers, additional strong blocker condition heavily burdens the SFDR of the ADC. Therefore, when comparing ADCs, all the functions of the data-acquisition channel should be properly addressed.

Note that both the sampled noise and thermal noise are spread uniformly over the Nyquist band as shown in Fig. 4.10. Therefore, an important conclusion can be drawn. If oversampled, the in-band noise power spectral density including the quantization noise is reduced by −3 dB per oversampling by two. This gives a clear incentive to choose the oversampling ADC over the Nyquist-rate ADC unless limited by the operational speed. The digital output of the DSM is low-resolution bits coming out at oversampling rates, and it should be filtered and decimated down to the Nyquist-rate digitally, which inevitably results in a long latency if linear-phase digital finite impulse response (FIR) filter is used. On the other hand, the Nyquist-rate ADC requires a significant amount of analog AA filtering, and sample the wideband output noise of the previous stage. In fact, oversampling opens up the possibility of performing AAF and AGC functions more accurately in the digital domain.

For fair comparison, two ADC systems of the open-loop SAR and the closed-loop second-order CT DSM are sketched in Fig. 4.11, which perform the same

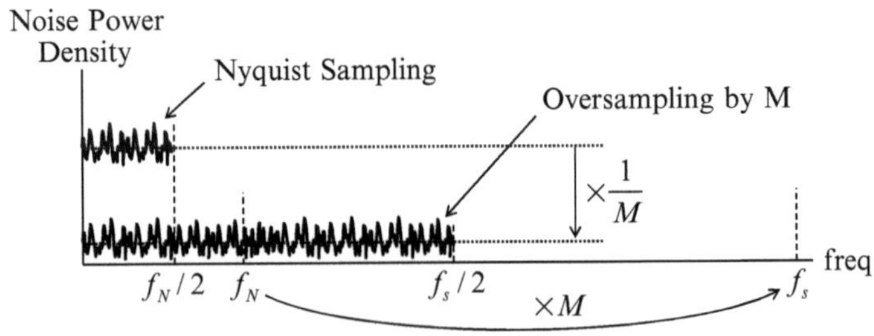

Fig. 4.10 In-band noise spectral densities

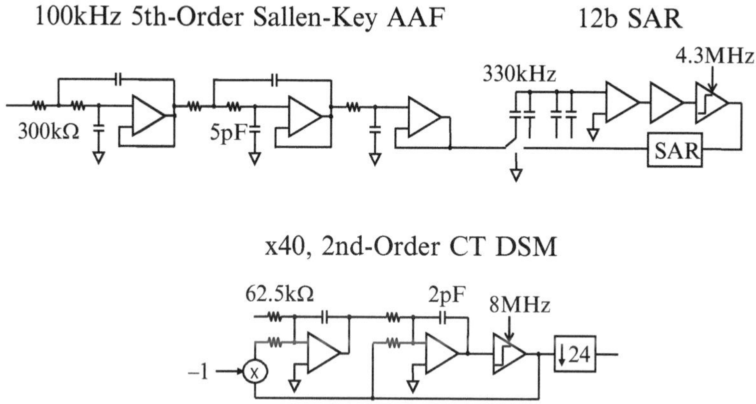

Fig. 4.11 SAR and CT DSM for 12b, 100 kS/s ADC

equivalent function from input to output. Both yield 12b outputs at 100 kS/s. First note the sampling rates of 4.3 and 8 MS/s are not significantly different. The AAF requirement for SAR prohibits sampling at the Nyquist-rate of 200 kHz. With a fifth-order Butterworth AAF, the sampling rate can be lowered to 330 kHz. To resolve 12b, SAR needs to sample 13 times faster at 4.3 MS/s. For CT DSM, the 12b resolution is obtained by oversampling 40 times at 8 MS/s using only a 1b second-order modulator. Third-order modulators can sample at even lower rates with no AAF than SAR which requires fifth-order AAF.

One obvious difference is that SAR comparator should make all 12-bit decisions with fine 12b resolution while CT DSM can make only 1b decisions. If the comparator is not reset as in the SAR operation, they also suffer non-Gaussian comparator error due to the comparator hysteresis. Open-loop SAR works only with a large Volt-level input. On the other hand, CT DSM is operable at low supply with an input at least −20 dB smaller than the SAR. SAR requires comparators to resolve extra 3 more bits for the acceptable BER. To get $N+3$ bits of comparator resolution, resettable preamplifiers should have a small-signal gain of 2^N and a bandwidth wider than the Nyquist bandwidth by $\ln(2^N)$, which are 72 dB and $8.3(f_s/2)$, respectively if $N=12$ as discussed in Chap. 1. Typically such wideband open-loop amplifiers can be implemented only by cascading multiple stages. In the CT DSM, only two integrator stages are needed.

The AAF requirement is far more serious in the SAR since it needs a fifth-order Butterworth filter to pass the signal below 100 kHz but to attenuate the spectrum to be aliased at 330 kHz. Even for the Sallen Key filter, three unity-gain buffers are used. Active buffers are power consuming and noisy. The R and C values are large to make this low cut-off frequency. They are very nonlinear if implemented using the simple source follower buffer. Furthermore, the phase nonlinearity due to the non-uniform group delay needs also to be equalized at the system level.

In the open-loop SAR, capacitors should be matched to 12b level while no matching is required in the CT DSM. Thus every design parameters should be controlled to meet the absolute accuracy in the open-loop SAR. On the other hand, in the CT DSM, the open-loop transfer function suppresses the spectrum at the sampling frequency to a greater degree than the low-frequency signal band, and no AAF is even required. Since the loop filter works in feedback, the linearity of the loop filter is improved by the loop gain. Furthermore, the R and C values are much smaller since the bandwidth of the loop filter is $f_s/2\pi$, which is much wider than that of the SAR.

The current trend is that modern DSP-based systems perform analog functions such as AGC, AA, and channel filtering all digitally. They require high SFDR for digital blocker filtering in RF receivers, and for digital AGC in imagers and medical devices. CMOS SARs have been used for low-resolution applications, but it is challenging to implement their purely CMOS versions for resolution higher than 10b without losing the benefits such as simplicity and low power.

4.3 Incremental DSM with DC Input

The oversampling DSM is the most desirable among ADC architectures for high resolution mainly due to its simplicity and relaxed AAF requirement. Thus it has been replacing most Nyquist-rate ADCs starting from the low end of the spectrum such as for instrumentation, voice, audio, video, and wireless applications. As noise is spread over the oversampled bandwidth, it achieves a wide dynamic range without stringent capacitor matching and high opamp gain requirements. Its CT version (CT DSM) is an ideal choice when quantizing the baseband spectrum in digital wireless receivers. It requires no AAF and operates with much lower noise (−20 dB) than any Nyquist-rate ADCs can afford, and enables digital AAF, channel filtering, and even AGC functions.

The only lingering question has been the latency due to the digital post-processing, which is troublesome in certain applications though it is not an issue in general. SAR ADCs with 18b at low MS/s throughput rates have been widely used in precision data-acquisition instruments for digital imaging, medical, and industrial uses [15]. Compared to other pipeline and oversampling DSMs, they require neither accurate residue nor long latency. However, critical requirements such as capacitor matching, stress-induced comparator offset hysteresis, and pre-amp gain-bandwidth product are too demanding to meet with CMOS. Capacitors should be factory trimmed or calibrated. Comparator should make decisions with full resolution while input S/H and high-gain open-loop preamp are working at much higher sampling rates than the throughput rate. To make CMOS SARs work, many solutions have been suggested to be effective: (1) Resetting comparators before every bit decision, (2) digitally correcting comparator errors as in the pipeline, and (3) using hysteresis-free BiCMOS for preamplification. That is, CMOS SARs are acceptable for low resolution, but not suitable for high resolution. When SAR is modified like a pipeline with an interstage residue amplifier, a two-step SAR offers 18b even at over 10 MS/s [16].

On the other hand, the quantized feedback makes DSMs achieve high resolution with no accurate capacitor matching and high opamp gain. The long latency period can be shortened by periodically resetting modulators and averaging the digital outputs until reset again. If DSM is operated in this transient mode as a digital decimator or *T/H*, it is called incremental DSM (IDSM). Such operation demands that the input be narrow-banded and stay constant during the tracking period. As a result, they still sample at much higher rates than necessary, and their applications are limited to very low-frequency systems with high oversampling ratios [17]. High-throughput IDSMs with low oversampling ratios still decimate digitally, and operate like hybrid two-step or pipeline together with other SAR or recycling ADC [18, 19].

4.3.1 IDSM with Input S/H

When operated with sampled DC input, DSM works no differently from other Nyquist-rate ADCs, and offers the same high-resolution performance as is possible using only trimmed or calibrated SAR. Latency is the time delay from when the input is sampled to when the digital output is valid. The SAR output is delayed by minimum $N+1$ clock periods for sampling and N-bit decisions while the pipeline delay is just the number of pipelined stages. Parallel ADC like flash has only one clock delay for sampling and decision while serial slope-type ADC requires 2^N clocks. DSM is more like a serial tracking ADC, and achieves high resolution by digital decimation and filtering. The former is to lower the sampling rate by notching out the noise spectrum around the frequency to be decimated to, and the latter is to filter out the noise and quantization error outside the signal band before the final decimation. The latter low-pass and final decimation filter requires a long latency period accordingly, and the situation gets worse as its phase response should be linear. The linear-phase delay of the in-band signal is achieved only when the in-band group delay is kept constant. That is, all frequency components should come out of the filter after an equal fixed time delay.

Digital infinite impulse response (IIR) filter can be designed to be linear phase only within the signal band. Although FIR filter provides the inherently linear-phase delay characteristic, it exhibits prohibitively long group delay as the number of taps grows. To avoid the latency, IDSM operates without the final low-pass and decimation filter which offers the desirable feature of further suppressing the noise and quantization error. Due to the lack of this predecimation filter, the IDSM is somewhat inferior in noise performance to the normal DSM.

There are two ways to apply the input to IDSMs as shown in Fig. 4.12. Using DSM, the band-limited signal can be sampled and reconstructed at discrete times. One Nyquist sampling interval is expanded with high oversampling points marked in between with the dashed lines. The input voltage at the throughput rate is marked with black dots. In both cases, the latency is shortened by operating the DSM within the throughput-rate interval. On the left side, the normal DSM is operated like

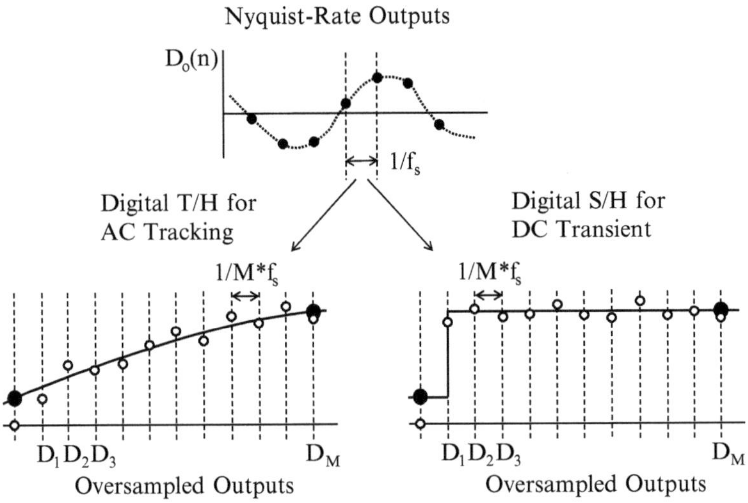

Fig. 4.12 Signals in normal and incremental DSMs

digital *T/H*. It starts to take the AC input from the reset state at the beginning, and tracks the input by oversampling M times. The decimated throughput-rate digital output is obtained before reset again for new input sampling. Accuracy is lost in this digital decimation as the input varies, and no final filter stage is used. Therefore, it requires very high oversampling ratio or the input should be narrow-banded. On the contrary, if the input is frozen by an analog *S/H* during the sampling interval as shown on the right side, the DSM doesn't need to track the AC input. In the transient DC mode, the input *S/H* is analogous to the decimation performed in the analog domain. That is, the already-sampled DC output of the throughput-rate *S/H* is resampled repeatedly by *M* times. It works like a digital *S/H* that only needs to settle to the constant DC output.

If an incremental input voltage is applied to the DSM in the form of the step input, the DSM feedback loop settles to its digital final value with characteristic time constants of multiple poles starting from the initialized state. Since the modulator is reset, it becomes the regular pulse-coded modulation (PCM). The unity-gain-bandwidth of the DSM loop is normally $1/2\pi T = f_s/2\pi$ since the integrator frequency response is $1/j\omega T$. That is, the loop settles with a time constant of T or $1/f_s$. In the simple case of a single dominant pole response, the DSM output approaches the final value as follows.

$$D_o(m) = D_o\big|_{\text{Final}} \times \left(1 - e^{-m}\right), \tag{4.2}$$

where $D_o(m)$ is the oversampled digital output that can be defined in discrete times.

The input *S/H* facilitates the Nyquist-rate data acquisition since the loop is closed in unity-gain feedback, and the IDSM works like a digital *S/H* [20]. This implies that the DSM digital output settles to the final value with *N*-bit resolution

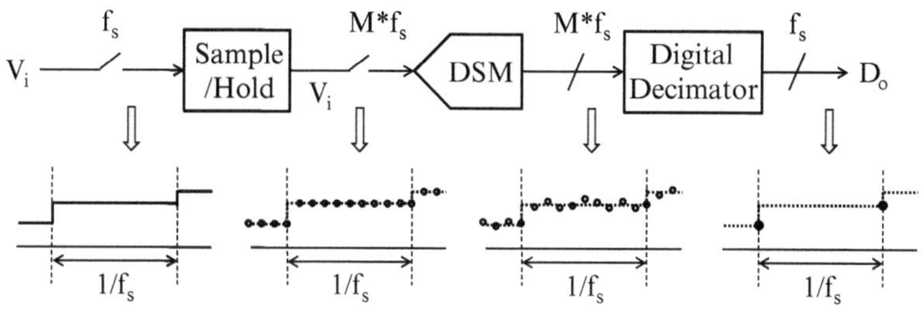

Fig. 4.13 Concept of IDSM operation with DC input

after the initial $\ln(2^N)$ clocks. For example, the transient output settles with 10 and 16b resolution like an analog sample/hold amplifier after about 7 and 11 clocks, respectively. This digital *S/H* output after fully settled yields the DC samples during the remaining period of one throughput-rate sampling interval. If oversampled by 32 times, $(32 - 11) = 21$ digital samples of one DC output are available with 16b accuracy.

The operational concept of IDSM with input *S/H* is shown in Fig. 4.13, where the last steep-slope digital low-pass filter with very low cut-off frequency is missing. It is also a causal system as all integrators are reset at the throughput rate. Therefore, the steady-state filtering concept doesn't apply to this transient system, and only the time-domain discrete-time processing is valid. That is, during the sampling period, only M digital samples $D_o(n)$ for $n = 1, 2, \ldots, M$ are available for the final DC estimation. So in the time-domain, the digital output data are the stream of the quantized DC values of the *S/H* output $V_i|_{S/H}$ with three non-ideal error terms.

$$D_o(i) = DV_i|_{S/H} + \sigma_q + \sigma_{tr} \pm \sqrt{\sigma_n^2}, \tag{4.3}$$

where note that the magnitudes of non-ideal error variances are added for explanation purpose only. Three are contributed by the quantization error, the initial transient error, and the random noise. Except for the last random noise variance, the first two errors are deterministic, and increases the BER of the ADC. Since the bit error can be considered as the quantizer error larger than the quantization step, the deterministic errors should be reduced below the quantization level.

The quantization error spectrum is high-pass shaped in the modulator if resonator poles are not used in the loop filter. For example, a fifth-order modulator shapes it by s^5, and the same-order integrator can recover it by $1/s^5$. Therefore, one higher-order slope of $1/s^6$ is sufficient to remove the quantization noise. The transient error occurs since the output of the modulator cannot follow the step input function instantly. In the time-domain, only the ideal transient step or a single-pole response is linear. Therefore, the DC estimation boils down to how to estimate the transient output of the digital *S/H* that settles with multiple poles and zeros. On the last random noise, the noise matched filter for DC is a low-pass filter with extremely low cut-off frequency, but the long latency of the FIR filter prohibits its use.

4.3.2 Cascaded Integrators for DC Estimation

In digital signal processing, it is well known that the time-domain integrate-and-dump function is a good matched filter for the impulse symbol. It is known to reconstruct the impulse symbol while rejecting the high-frequency noise. Since integrator is DC-unstable, the integrated output is divided by the total number of data samples to make an averaging filter such as SINC. SINC is the gated running sum filter that gives the comb filtering function with zeros at the multiples of the sum frequency, but such steady-state filtering is not valid in the DC transient mode. The integrate-and-dump performs the same time-domain averaging function with a limited number of samples. However, the high-order integrator is not a good comb filter, and the imperfect decimation raises the aliased noise. Five integrators are cascaded to average output data as shown in Fig. 4.14 for $M = 30$. Also note that the high-order integrator weights the earlier data more heavily than the later ones.

The IDSM output has been commonly estimated using this high-order integrator from the reset point. The intention is to separate the almost DC spectrum from any AC error and noise by averaging. That is, the quantization noise-shaping matters little as long as the constant DC can be separated from the remaining quantization error and noise. However, in IDSM with input analog *S*/*H*, the output is not constant DC in steady state but rather time-varying in transient. Since the higher-order integrator averages data quickly with a fewer number of samples, they use the earlier samples repeatedly to get the average quickly with fewer samples. This implies that the random noise is not averaged out since all the independent random variables are not equally weighted. That is, it is not effective in reducing

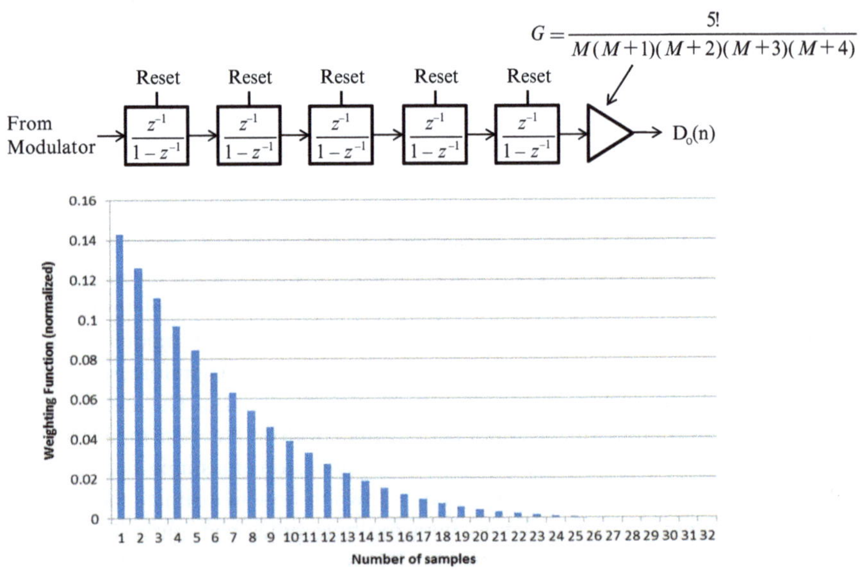

Fig. 4.14 Cascaded integrators and effective data weighting

the higher-order shaped quantization error, but not the random noise variance. Therefore, the integration should be delayed to avoid the initial transient error until it is cleared from the modulator output.

Other possible alternatives are to use IIR filters with short latency since there is no linear-phase requirement for the DC input, or to adaptively estimate the DC output minimizing the mean square error (MSE) such as Kalman filter in the Stochastic random process. If the true mean of the data can be estimated independently of the error variance using an MSE, it is possible to achieve finer resolution than N bits. More elaborate DC estimation algorithms operating the high-order integrate-and-dump and the MSE in parallel can facilitate the convergence of the MSE with fewer samples than averaging filters though the concept has yet to be demonstrated.

4.3.3 Switched-Capacitor Charge Injector for Input S/H

The performance of the IDSM entirely depends on the availability of an accurate frozen *S/H* input in order to sample it repeatedly during the sampling period. Designing such high-resolution sample/hold is one of the most challenging tasks in analog designs. However, the input stage of the DSM is an integrator that takes the difference (Δ) of the input and the feedback DAC output, and integrates (Σ) the error for feedback. Therefore, it is possible to use a constant charge as an input as shown in Fig. 4.15.

The input can be applied in either no-return-to-zero (NRZ) or return-to-zero (RZ) forms. The charge input to the integrator is equivalent to adding a linear voltage-to-current converter to the *S/H*. The switched-capacitor charge injector (SCCI) converts the voltage input into charge and injects it into the integrator. The constant charge is injected by M times in the NRZ format while less than

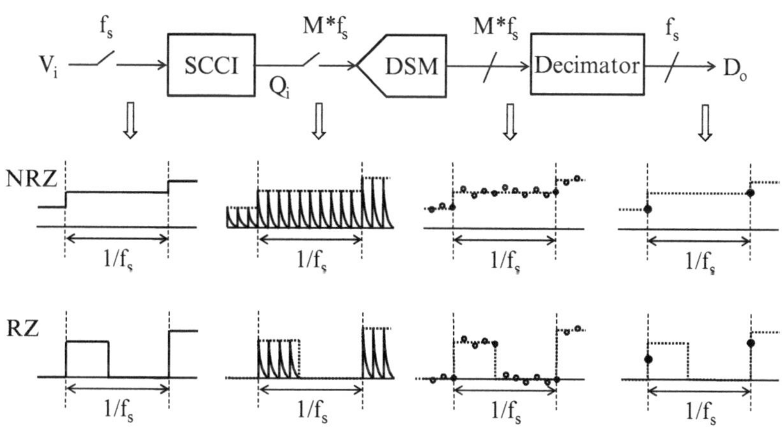

Fig. 4.15 NRZ and RZ charge inputs

M times in the RZ format. The analog input is not band-limited as it is sampled on the input capacitors, and the output data stream is the digitally oversampled value. So the input charge is injected into the integrator in discrete times. Therefore, the total input charge is related to the input by

$$Q_{\mathrm{i}} = (M \times C) \times V_{\mathrm{i}}. \tag{4.4}$$

The factor M results because the constant voltage input is equivalent to integrating the input charge of CV_{i} by M times. The total charge input Q_{i} is injected into the integrator during the whole sampling period. The advantage of the input charge injection is evident since it gets rid of the input S/H, and greatly simplifies the implementation. It is true that the total amount of charge injected is constant and independent of the matching accuracy of M sampling capacitors. However, the capacitor mismatch contributes to the noise by modulating the input magnitude, which otherwise stays constant over the period. When the DSM output is filtered or estimated, the capacitor mismatch noise is partly filtered, but the remaining noise may degrade the BER.

The weighting effect of the high-order integrator opens up a possibility of the lumped input charge injection using the RZ format at the beginning of the sampling period. For example, assume that 30 capacitors are needed to inject charges 30 times. If the SCCI is made of 10 capacitors, the input charge is integrated only 10 times, and the input is grounded for the remaining 20 clock cycles. Since the duty of the charge injection cycle is 1/3, the average input is in effect scaled down by the same ratio. However, the high-order integrator weights the earlier charge inputs more heavily than the later ones, and the actual signal is attenuated only by 0.74 instead of 1/3. This loss can be made up by sizing up the input capacitor. That is, a less number of capacitors can be used, and the longer tracking time can be used for data acquisition. The input charge is still modulated by the capacitor mismatch, but in the RZ input case, the total injected charge stays constant independently of the charge shape and capacitor mismatch since the charge integration is finished early during the digital sampling period.

4.3.4 *Initial Transient Error*

To avoid the integrator overloading, the input is often directly fed to the comparator input while letting the error go through the integrator path as shown in Fig. 4.16.

This feedforward modulator has been a staple in IDSMs since the signal doesn't need to go through the integrator path but only the error does [21]. However, the output data still suffer two errors. One is the quantization error, and the other is the initialization error. The former has a zero-mean with a variance, and can be handled like random noise. The latter contributes to the transient distortion and noise, which result from the difference between the actual transient waveform and the ideal step or the exponentially settling waveform with a single pole. Since the input is

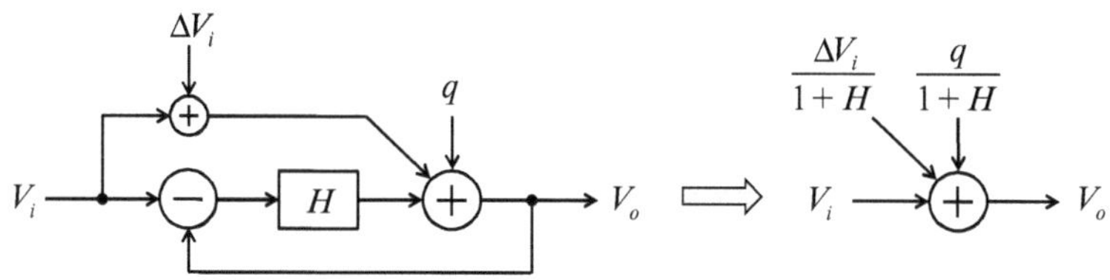

Fig. 4.16 Model for modulator with input feedforward

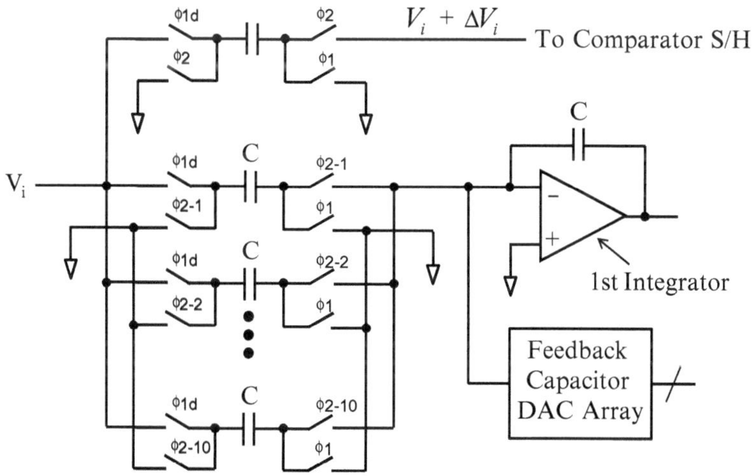

Fig. 4.17 SCCI input and the feedforward path

forwarded to the comparator input, the signal transfer function (STF) becomes all-pass, and the difference between two paths is injected into the quantizer as an error.

$$V_o = V_i + \frac{\Delta V_i}{1+H} + \frac{q}{1+H}, \tag{4.5}$$

where ΔV_i and q are the path mismatch and the quantization error, respectively. Note that both the initialization and quantization errors are high-pass shaped. This mismatch error should be suppressed by the loop gain to be smaller than the quantization error. With AC input, the error is injected into the loop M times, but with DC input, once at the beginning like a step error function. It is the sampled error affected by input frequency, clock skew, kT/C noise, clock feedthrough, switch nonlinearity, etc. Therefore, the IDSM output settles to the correct value after this initial transient error disappears.

The SCCI is a switched-capacitor (SC) integrator with a bank of multiple input sampling capacitors as shown in Fig. 4.17. The lumped charge is injected only at the beginning of the oversampling period [20]. The initial transient error occurs during the brief period from when the step mismatch error ΔV_i is forwarded to the comparator input to until it is suppressed by the loop gain together.

4.4 LMS-Based Adaptive Error Feedback

From the early days of old discrete electronics, engineers have been grappling with inaccurate components when setting circuit parameters. The task has been carried out in the form of adjusting trimpots and varactors on circuit boards. More elaborate factory trimming on chips has been common to manufacture precision analog products such as voltage reference and high-resolution ADC/DAC. All electronic trimming systems require three distinctive functions of measuring, detecting, and correcting errors. It is the same working principle as the first-order DSM loop. The first error measurement or detection performs the delta (Δ) function while the second error polarity decision is the sigma (Σ) process. The final error adjustment or trimming completes the error feedback. This error feedback algorithm can be implemented in either continuous or discrete times, and can even include the human intelligence inside the loop. When it is electronically automated, it is called self-calibration and self-trimming.

An early example of the automatic line buildout equalizer (ALBO) for telephony is sketched in Fig. 4.18. It is to equalize the channel frequency response so that the length of the line may not affect the voice-band transmission and reception. That is, the variable filter bandwidth is trimmed using the feedback control signal based on the comparison (Δ) of the peak value with the ideal value and the error decision (Σ) to apply negative feedback. Since this parameter trimming is done in slow discrete times like the slow servo mechanism, it can be called zero-forcing servo feedback. These days, all modern analog/digital systems are built on some kinds of advanced electronic feedback concepts from the system level such as adaptive decision-directed equalizer and echo canceller to DC offset and gain trimming based on the least mean square error (LMS) algorithm.

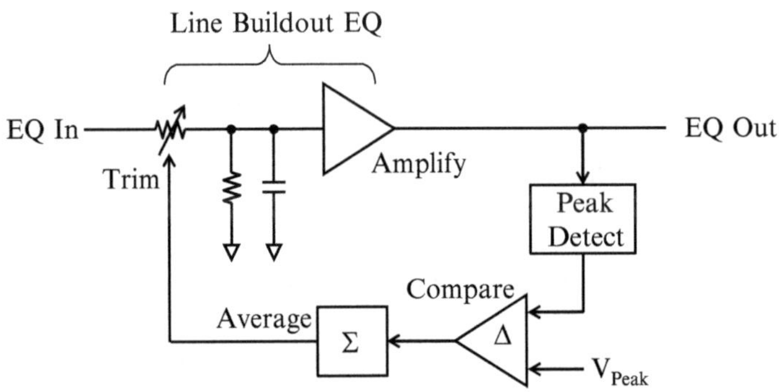

Fig. 4.18 Early self-trimming example of ALBO

4.4.1 Loop filter for Stability

The error detector operates in two different ways. One is to measure the error value itself, and the other is to sense the polarity of the error only. The former is just a highly oversampling DSM to be used for digital self-calibration while the latter embodies the electronic self-trimming concept. The former can be digitally implemented to make more complicated adaptive systems while the latter performs an automated analog trimming procedure for individual parameters though digital memory holding error values can be updated incrementally one bit by one bit. Such feedbacks for performance enhancement include long digital averaging cycles (>1000) and even operate in discrete times. It uses the same integrate-and-dump concept common in digital signal processing. Averaging a certain number of data in time gives the standard SINC comb filter function in the frequency domain with the frequency response of $\sin x/x$. With the SINC filter inside the loop, the feedback loop is stable with a good phase margin close to 90° like it has a dominant pole at a frequency much lower than the unity loop-gain frequency. The gain of the SINC filter drops with the $1/s$ slope, and as in the feedback amplifier, the stability condition that the dominant pole frequency is lower than the unity loop-gain frequency by more than the loop gain applies. The negative servo feedback becomes in effect a high-pass filter with a very low cut-off frequency at DC. If the parameter to calibrate or trim is constant, it gets into the steady state with no DC wandering problem.

Figure 4.19 illustrates an example of the DC correction loop by feedback in *I*/*Q* down-conversion wireless receivers. The offsets in the *I*/*Q* paths degrade the signal constellation due to ISI. First, the estimate of offset is subtracted, and the residual offset is integrated digitally. The DAC register holding current offset is updated accordingly. This slow servo loop stabilizes the DC fluctuation in each *I*/*Q* path. It is in effect a high-pass filtering the offset with a low cut-off frequency, and it suffers greatly from the DC wander as the low-pass signal spectrum is also centered at DC. Fortunately, all digital communications transceivers send or receive data in packets with finite lengths always with head and tail. Therefore, discrete-time offset

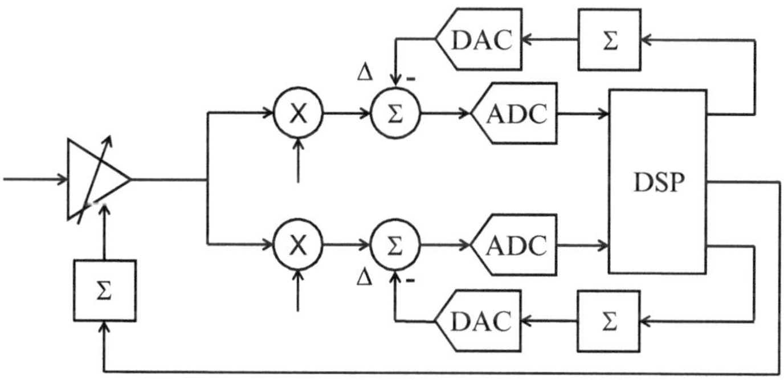

Fig. 4.19 Offset cancellation in *I*/*Q* down-conversion

feedback is possible using an offset DAC which is updated and held constant during the period of the data packet. However, the recent trend is towards the digital software-definable radio that can correct offset and perform AGC both in the digital domain using wide dynamic range *I*/*Q* ADCs.

4.4.2 *Self-Calibration vs. Self-Trimming*

There are two general ways to correct a circuit parameter error. Common self-calibration can be done digitally, but it is the open-loop feedforward correction often done digitally. It cancels the error numerically or using look-up table in real time at the output. However, the effectiveness of the self-calibration depends on the measurement accuracy. That is, the inaccuracy in the error measurement is not corrected until the next calibration cycle. On the other hand, self-trimming can eliminate the error continuously as long as the feedback is engaged. In a sense, the self-calibration is a volatile method while the self-trimming is a nonvolatile one. The adaptive self-trimming can be implemented either in the analog or digital domain, and error is corrected at its very source like the trimmed part works perfectly. Discrete-time feedback can encircle extra elements such as human, sensor, and mechanical part in the loop as long as a long time constant given by digital adder or integrator is counted in the loop for stability. Two very common open-loop self-calibration and closed-loop adaptive self-trimming schemes are shown in Fig. 4.20.

The self-calibration often called digital calibration and/or digital enhancement/aiding is an error feedforward system based on the multi-bit quantization of the absolute error, which requires long error measurement cycles and complicated numerical error calculations during the normal operation [22–24]. On the contrary, the self-trimming is an adaptive error feedback system based on the oversampling 1b DSM for error polarity detection [25–28].

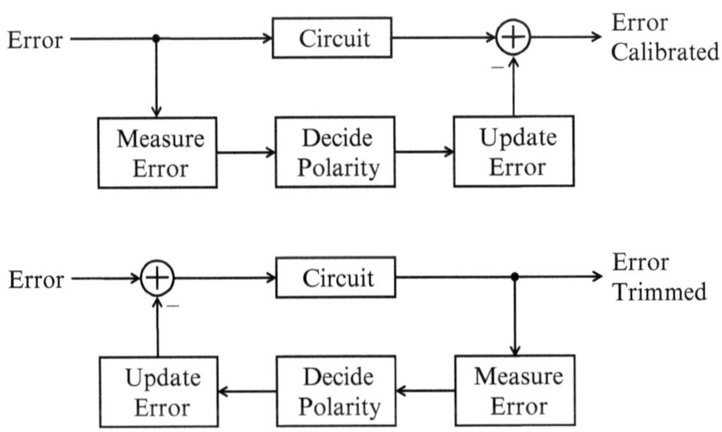

Fig. 4.20 Self-calibration vs. self-trimming

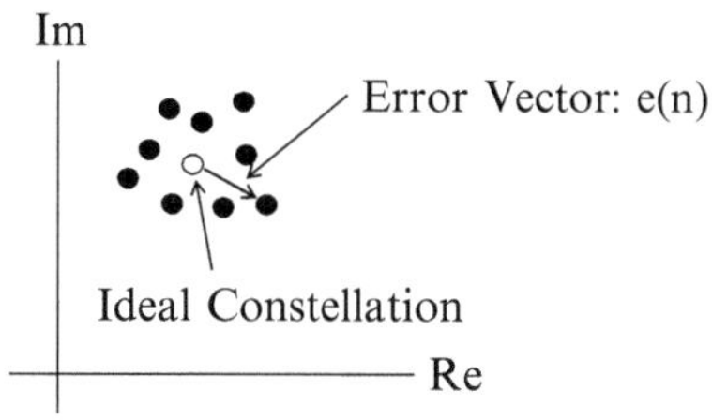

Fig. 4.21 Decision-directed equalization for symbol error

Once the error is acquired in self-trimming, the error update time is short due to its tracking nature. Any parameter in question can be trimmed both in the analog and digital domains. That is, the issue boils down to the multi-bit quantizer vs. -single-bit quantizer in the error measurement. In fact, the self-trimming is an extension of the classic analog design with some digital aiding in the error polarity detection. However, high-resolution analog designs went digital and disappeared rapidly in too much a hurry, ending up with all digital self-calibration for digital aiding and enhancement. The self-trimming moves high-resolution analog designs back to the analog domain.

LMS algorithm originates from the filter coefficient adaptation for the equalizer that warrants the MSE condition of the received signal as shown in Fig. 4.21, where one received constellation is sketched. In two-dimensional plane, received data symbols are scattered around the ideal symbol. The individual error vector $e(n)$ is defined as the distance from the ideal symbol, and the error power is the sum of $e^2(n)$. Since the received signal should be ideal, the decision-directed equalization algorithm to derive the MSE condition is well established in digital communication receivers. However, in electronics at analog circuit levels, it becomes a single-parameter adaptation. Most often than not, they are common circuit parameters such as gain, bandwidth, capacitor matching, resistor value, time constant, image, and fractional spur, etc. Therefore, if the LMS algorithm is applied to the single-parameter adaptation, it becomes the zero-forcing servo feedback, and feedback is based on the oversampling $\Delta\Sigma$ modulator concept.

4.4.3 Error Measurement by PN Dithering

The adaptive LMS algorithm can be implemented using a DC servo feedback loop that encircles both analog and digital domains. Pseudo-random (PN) dithering is commonly used, which facilitates the process of error correlation and signal de-correlation. An issue arises on foreground vs. background error measurement. In the former, the normal operation is interrupted to accommodate calibration cycles while in the latter, error can be PN-modulated, and embedded into the signal. During the normal operation, the PN-modulated error can be correlated out, and no separate error measurement cycles are required. The trend is obviously towards the background error measurement. So the resultant zero-forcing feedback works with

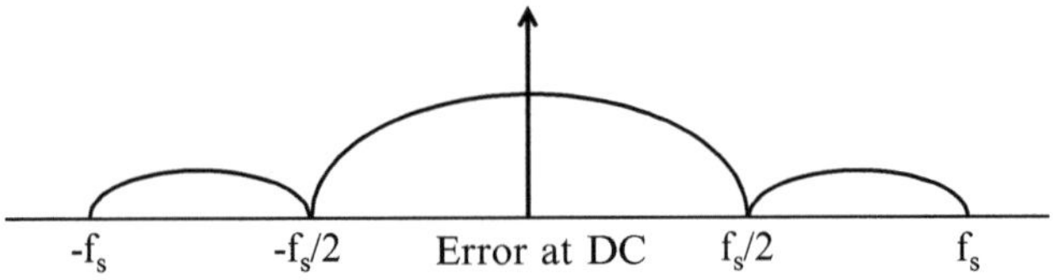

Fig. 4.22 Spectrum of digital PN sequence

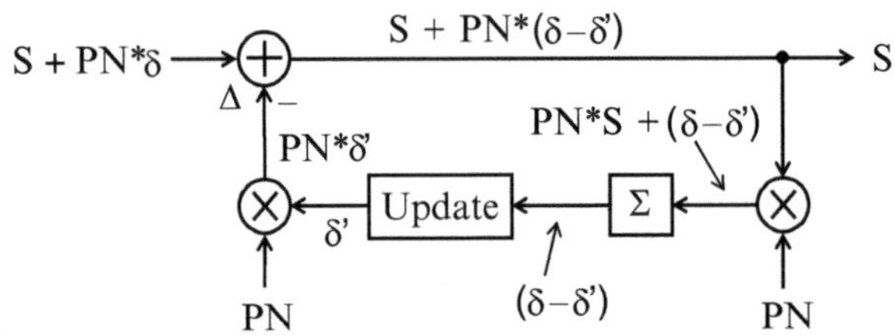

Fig. 4.23 Adaptive algorithm for PN-dithered single-parameter error

an oversampling error detector. The error polarity output of 1b can be integrated like the first-order 1b DSM.

The spectrum of the PN sequence is shown in Fig. 4.22. Multiplying by PN is the random binary pulse modulation called spread-spectrum used in GPS and CDMA. The PN sequence is a random time sequence of 1 and −1 (digital 1 and 0) with an equal probability for 1's and 0's over a long time period. If the error at DC is spread by the chipping clock oversampling by *M*, its spread spectral density drops by *M*. If it is despread using the same PN sequence, it grows back up with the process gain of *M*. The dispread DC error is obtained by the SINC filter, but the de-correlated signal is spread by the same chipping and filtered out.

The single-parameter adaptation facilitates the extraction of the MSE error as shown in Fig. 4.23. The error power polarity should be detected without measuring its absolute value so that the error feedback can work with only the polarity of the error. Assume that the small PN-modulated error PN*δ is embedded into the signal S. If its estimate PN*δ′ is subtracted (Δ) from the input, the output is the signal with a residual error PN*(δ−δ′). Now the output is correlated and averaged (Σ), only the error polarity of (δ−δ′) remains since the de-correlated signal PN*S is averaged out to be zero while $PN^2 = 1$. This polarity information is used to update the estimate to zero-force the error by matching δ=δ′. The single-parameter case is the degenerate case, and the LMS algorithm zero-forces the MSE rather than minimizing it.

4.5 LMS-Based Adaptive Servo Feedback Examples

Servo feedback can be applied to correct any DC parameters in analog systems such as component matching, offset, opamp gain and nonlinearity, time constant, image, and fractional tone. The feedback mechanism can be configured independently depending on how the system to calibrate operates. Self-calibration concepts have originated from the need to achieve high matching accuracy of capacitors to replace costly factory trimming procedures by electronic ones. However, fine lithography now offers inherently high speed and matching accuracy, and capacitor matching is no longer an important factor in data-converter designs except for the extreme cases of high-resolution Nyquist-rate ADCs such as SAR and pipeline over 16–18b levels.

4.5.1 Capacitor Self-Trimming

In analog sampled-data processing, the most fundamental block is the switched-capacitor amplifier, which produces a discrete-time output voltage with an exact gain set by the capacitor ratio. Such amplifiers are commonly used as a residue amplifier.

The two-capacitor basic multiply-by-2 amplifier is shown in Fig. 4.24. Assume that the opamp is ideal, and the input V_i is sampled on both the capacitor C and the other $C(1+\alpha)$ mismatched slightly during the sampling phase. The mismatch between them is given by α, which is much smaller than 1 ($\alpha \ll 1$). If the two capacitors are swapped in two different ways randomly depending on the PN sequence as shown, the mismatch error in the output is PN-modulated as follows.

$$V_o = (2 + \text{PN} \times \alpha)V_i - (1 + \text{PN} \times \alpha)V_{ref}. \tag{4.6}$$

Thus the core concept of the background calibration is to measure this PN-modulated error term by correlation using the same PN sequence. If this is used as a residue amplifier in the pipelined ADC, the output V_o can be correlated by the same PN sequence digitally, and accumulated. Since the PN sequence has an equal probability of being +1 and −1, the sum of n correlated outputs would approach the following value.

$$\sum \text{PN} \times V_o = \sum \text{PN} \times \left(2V_i - V_{ref}\right) + n \times \alpha \times (V_i - V_{ref}). \tag{4.7}$$

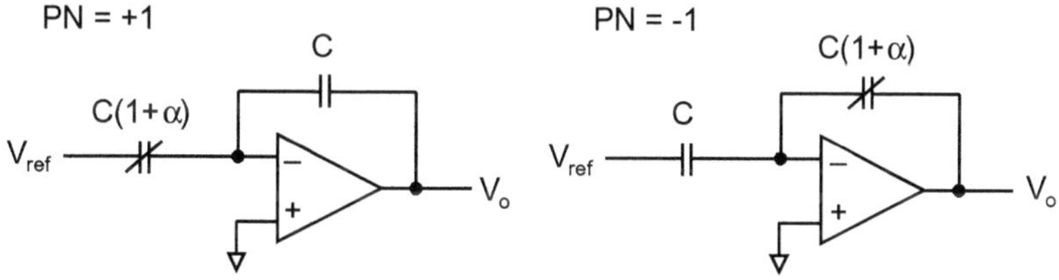

Fig. 4.24 Switched-capacitor amplifier with capacitor swapped

This implies that the uncorrelated residue output of the first term grows at the rate of $n^{1/2}$ while the error term grows linearly by n. For the measured error term to exceed the residue, the following condition should be met.

$$n > \frac{1}{\alpha^2}. \tag{4.8}$$

This often requires a very long cycle even to detect the error polarity only. The error polarity detection can be also implemented both in the analog and digital domains using the ΔΣ concept. If implemented using an analog first-order ΔΣ modulator, the constant part of the residue output can be subtracted to PN-correlate only the residual mismatch α term, thereby facilitating the error polarity detection by shortening the signal de-correlation time. The signal subtraction can be done easily in the input sampling network of the ΔΣ modulator. Although this analog signal subtraction is not perfect due to mismatches between the sampling capacitors, the residual error after subtraction becomes much smaller.

In pipelined ADCs, extra resolution is required for digital correction to eliminate the comparator error. The minimum is the tri-level MDAC with so-called 1.5b DAC. In the 1.5b multiply-by-2 residue amplifier using the two-capacitor MDAC, the residue output is $2V_i - b^*V_{ref}$ depending on the tri-level coarse bit decision b, which is +1, 0, and −1. When $b = 0$, random PN swapping or dithering of the capacitor mismatch α is possible without sacrificing the signal range. When $b = \pm 1$, it can still be PN-modulated using the constant V_{ref} added or subtracted during the normal operation as a dither. If two capacitors are swapped depending on a PN sequence, the error component appears as $V_{error} = \mathrm{PN}^*\alpha(b^*V_i - V_{ref})$. If this error is correlated with the same PN sequence, the output is $\alpha(b^*V_i - V_{ref})$. Although the signal-dependent term of $b^*\alpha V_i$ is averaged out to be zero, the polarity detection is far easier than digitizing the αV_{ref} term for self-calibration. The polarity of α to be detected is accurate and independent of $b^*\alpha V_i$ because it is always true that $(b^*V_i - V_{ref}) < 0$, no matter what the b^*V_i value is. Therefore, the capacitor mismatch can be trimmed progressively in the analog domain using a capacitor trim network. Once the α error is trimmed out, the residue output is free of any error resulting from the capacitor mismatch [28].

Since $\pm V_{ref}$ is subtracted depending on the coarse decision bit, b ($b = \pm 1$), the PN sequence to swap capacitors is multiplied by b to generate the correct PN-modulated ratio error in the residue output. The polarity of α determines whether to add or subtract an incremental amount of capacitance to set the correct ratio. Self-trimming continues until the PN-modulated αV_{ref} term disappears from the residue output. This zero-forcing feedback is limited by the polarity detection accuracy and the incremental capacitor step size. The lower bound of the minimum detectable ratio error is set by the de-correlated signal average if the residue output has a zero-mean uniform distribution.

Capacitor trimming is done using capacitive dividers as shown in Fig. 4.25. The 7b trim capacitor covers a 10b-level mismatch range for an effective 15b resolution. Initial trimming needs at most 2^6 cycles because it starts from the nominal value.

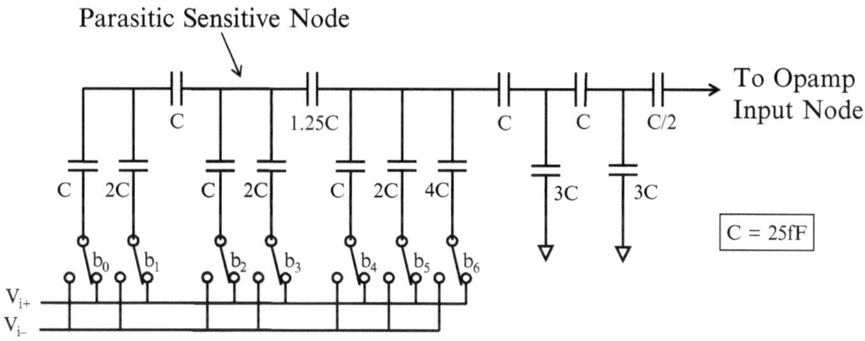

Fig. 4.25 A 7b capacitor trimming network

Although a binary search can reduce the initial trimming time, a thermometer search algorithm is simpler. If the MDAC unit capacitor is 400 fF, the trim capacitor can cover about the ±500aF range with an 8aF step with monotonicity. One LSB of the 7b trim capacitor can be updated after polarity detection. Since the trimming accuracy is limited by the accuracy of the polarity detection, the polarity detector offset should be nulled digitally first before the actual error detection begins.

4.5.2 *Self-Trimming DACs*

There are three kinds of common DACs used. They are made of resistor string, capacitor array, and transistor current source. The R-2R DAC is closer to the binary-weighted current DAC than to the resistor-string DAC. Thermometer-coded DACs based on the division by resistors or capacitors exhibit good DNL with monotonicity, but suffer poor INL. On the other hand, binary-weighted DACs exhibit good INL, but with poor matching tend to be non-monotonic with poor DNL. Resistor-string or capacitor-array DACs are usually used as a sub-DAC that provides reference levels for comparators in the ADC to make decisions. However, only current DACs offer dynamic performance to meet as stand-alone DACs. There are two criteria in the DAC measurement. They are static and dynamic performances. The static performance can be measured by static DNL and INL numbers while the dynamic performance is only measured by the transient distortion, which can degrade rapidly as the output frequency goes higher as shown in Fig. 4.26.

At low-input frequencies, SFDR, THD, or linearity is limited only by static DC nonlinearity, which result from mismatches among DAC elements and can be reduced by many ways such as using common-centroid, geometric averaging, trimming, shuffling, or calibration. At high-input frequencies, dynamic nonlinearity results from the transient distortion, and degrades SFDR [26]. The DAC transient step from one output to the other is shown in Fig. 4.27.

Fig. 4.26 SFDR vs. input frequency

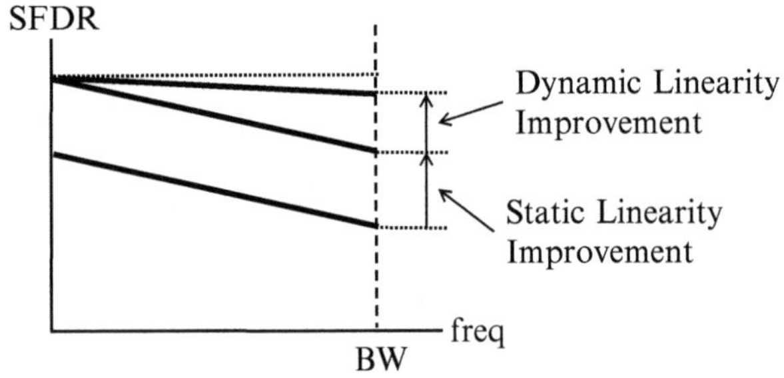

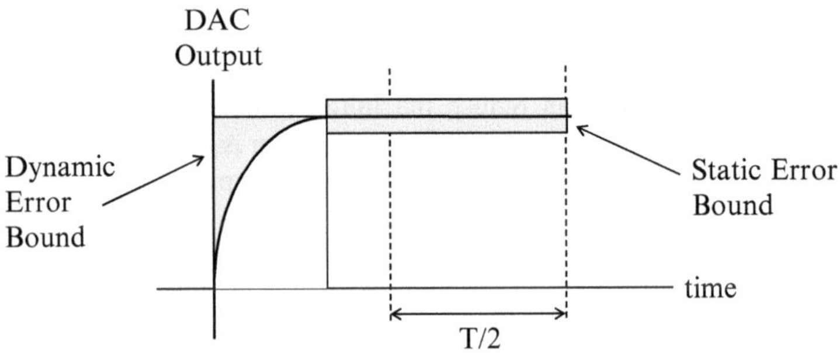

Fig. 4.27 Dynamic nonlinearity due to DAC transient error

The standard DAC output of the NRZ format is ideally a step function, but the DAC output is band-limited by network poles at the output node. The DAC transition occurs during the clock period T, and the DAC output should settle accurately. Two shaded areas mark the error bounds of the DAC output with static and dynamic errors. The static error is the error in the final voltage the DAC is settling to while the dynamic error is the transient error when the output is in transition. Each pole contributes to the transient output error with its own time constant. Therefore, the stand-alone DAC design demands that the DAC output should settle with a single dominant pole. Dynamic errors are fundamental in stand-alone DACs and continuous-time $\Delta\Sigma$ modulators. However, since the signal is already sampled in sub-ADCs, the DAC error is mostly the static error as long as the DAC output settles with enough accuracy within the clock period. Two examples of linear and nonlinear DAC settling cases are shown in Fig. 4.28.

The exponential settling with one-pole time constant τ is linear since the area of error h/τ is proportional to the step height h. On the other hand, the nonlinear settling case on the right side slews with a slew rate of S, and the area of error $h^2/2S$ is proportional to the square of the step height h. For slewing DACs to be linear, the slew rate should be much higher than the slope of the maximum sinusoidal voltage change of ωV_o. Therefore, there are only two ideal cases of the linear DAC transient. One is the ideal step, and the other is the single-pole exponential settling. Any DAC transient waveforms other than the ideal two are nonlinear.

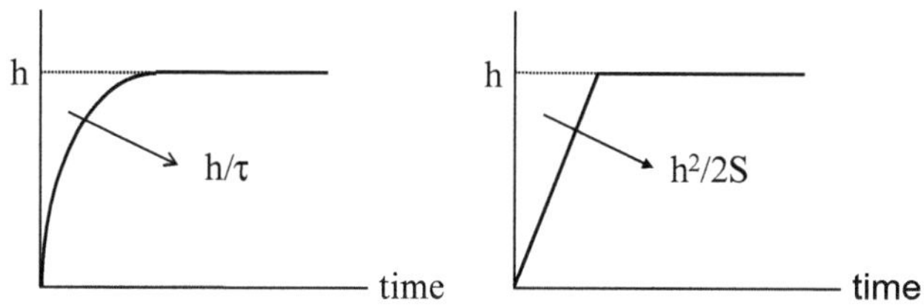

Fig. 4.28 Two examples of linear and nonlinear DAC settling cases

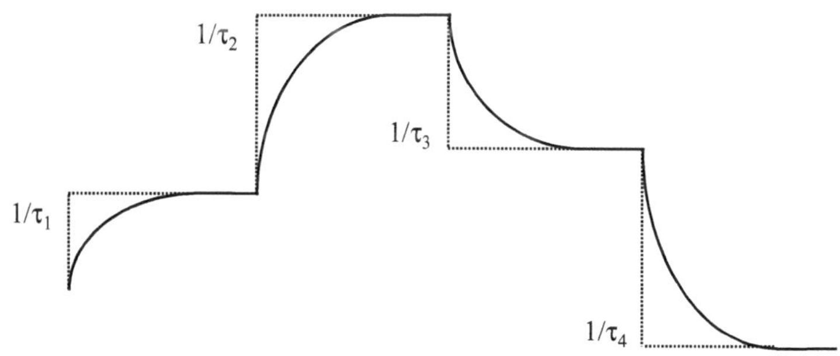

Fig. 4.29 Code-dependent DAC settling

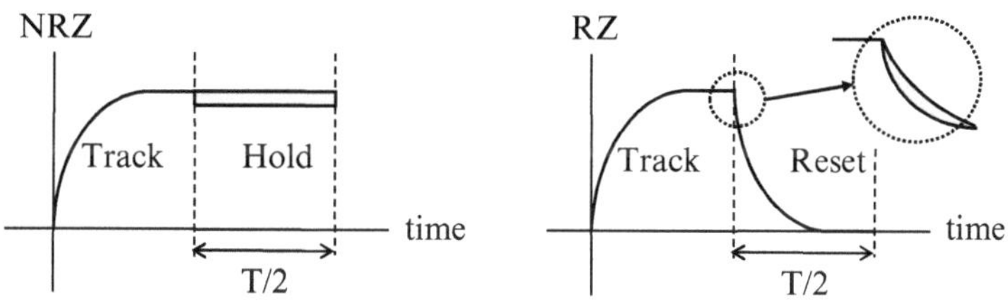

Fig. 4.30 NRZ and RZ formats of the DAC output

Figure 4.29 shows the transient DAC output example with code-dependent time constants. The DAC output time constant varies depending on how many DAC elements are switched to the output per digital codes. For high SFDR DAC designs, settling time constants should be kept constant for all codes. Even multiple poles of the DAC affect dynamic linearity. In addition, timing errors such as jitter, clock skew, and glitch also affect dynamic linearity. Glitch occurs when MSB in binary-weighted DAC is switched a little early or late, but the glitch energy can be minimized by using thermometer-coded DACs. There are two DAC output formats: NRZ and RZ. The former is more like the track and hold waveform while the latter is the track and reset waveform [26]. Their transient errors are illustrated in Fig. 4.30.

The common NRZ output is more sensitive to the static hold error. It is challenging to freeze analog voltages in time, and it is inevitable to have the hold error from charge injection and switch feedthrough. In the NRZ case, the hold error stays constant during the half-clock hold period. The square area is often used as a measure of nonlinearity error in V*sec unit as in the glitch.

$$\text{Error} = V_{\text{hold}} \times \frac{T}{2}. \tag{4.9}$$

It causes nonlinearity if it is not proportional to the output step. On the other hand, in the RZ case, the static hold error dies out exponentially as the output is reset.

$$\text{Error} = V_{\text{hold}} \times \tau\left(1 - e^{-\frac{T}{2\tau}}\right). \tag{4.10}$$

That is, the error decreases exponentially, and the exponential error is also linear. The advantage of the RZ format is that it is less sensitive to the word clock jitter and the passband droop is less severe than in the NRZ case. Furthermore, the high-frequency band is repeated at twice as high frequencies, and the final smoothing filter requirement is alleviated. For the reason, they are often used for charge injection circuits. However, the dynamic error power is doubled due to the rising and falling transitions unless it is a single-pole exponential settling.

To sum up, stand-alone DACs suffer from both static and dynamic nonlinearities. So called self-trimming or self-calibration is not to correct dynamic DAC errors but static DC parameters such as mismatch and offset. There is no known way to electronically trim or calibrate dynamic DAC errors. They can be reduced only by fast clocking or by single-pole exponential settling. However, for DACs used in ADCs, only static DAC errors count since the DAC in the sub-ADC produces only the DC reference voltages for comparators to make decisions. Static DAC errors are fundamental in all data converters, but correctible. In oversampling DACs, static DAC errors can be commonly randomized, shifted out of band, and digitally filtered out. However, in Nyquist-rate DACs, it can be trimmed or calibrated either in the analog or digital domains. One degenerate case is the 1b DAC, which requires no trimming or calibration since the two-level DAC has no gain or mismatch error. The oversampling 1b $\Delta\Sigma$ modulator is free from static DAC errors, but loses its dynamic headroom since higher quantization noise overloads the integrator. Its continuous-time 1b $\Delta\Sigma$ modulator is more affected by the timing error due to the pulse width and position jitters.

The feedback loop for other self-trimming examples is also the first-order $\Delta\Sigma$ error modulator as shown in Fig. 4.31 [25, 27]. It senses the resistor ratio mismatch and amplifier offset, and corrects them incrementally by small amounts using an up/down counter. The most common DAC trimming is the current DAC trimming. All practical stand-alone DACs are made of current-steering DACs with mostly MSBs thermometer-coded and LSBs binary-weighted. The monotonicity can be warranted if the next segment current in the thermometer-coded MSB array is

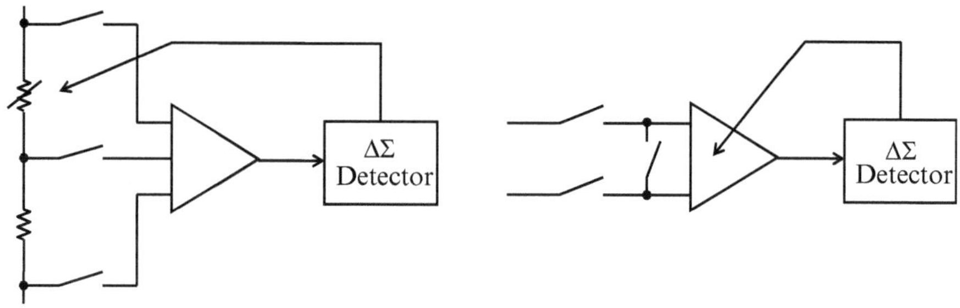

Fig. 4.31 Intuitive self-trimming for resistor ratio and offset

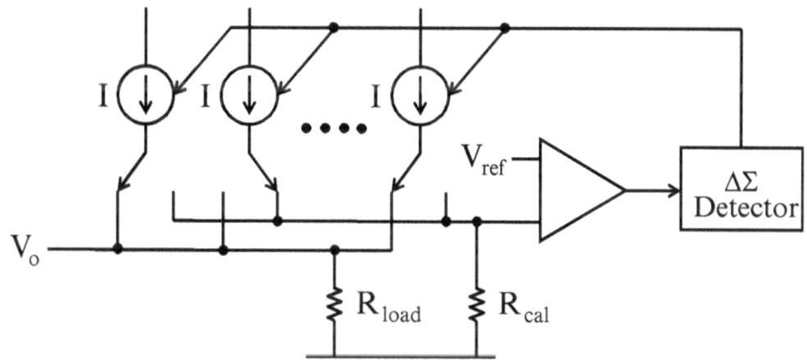

Fig. 4.32 Current source self-trimming concept

divided for the fine DAC current outputs. Since the DAC output is analog, only self-trimmed or self-calibrated MSB array can further improve static linearity unless the DAC output is predistorted digitally.

The self-trimming concept of the current source is straightforward as shown in Fig. 4.32 [26]. The current to be calibrated is directed into the calibration resistor R_{cal} so that the voltage of IR_{cal} can be developed. If it is compared to the fixed reference, the current mismatch can be detected. This detected error can be averaged to decide whether the current should be increased or decreased. Individual current sources can be trimmed in two ways. One is to add a parallel trim current DAC, and the other is to adjust the gate voltage of the current source using a voltage trim DAC, which is called current copier. The trim DAC output voltage is sampled on the capacitor tied to the gate so that the correct current can be sampled, but the copied current is volatile and needs updating.

Figure 4.33 shows highly linear differential RZ current-steering DAC featuring all performance enhancement techniques known to date. It eliminates errors related to code-dependent settling, slewing, *T*/*H*, glitch, and clock jitter. The down side is that the RZ format output is smaller, but the SINC error gets smaller too. The current DAC outputs are summed at the cascode node, and folded. They are dumped onto the load resistor. During the half-clock, the output is reset. The parasitic at the cascode node is code-dependent, but the output node settles approximately with a single *RC* time constant.

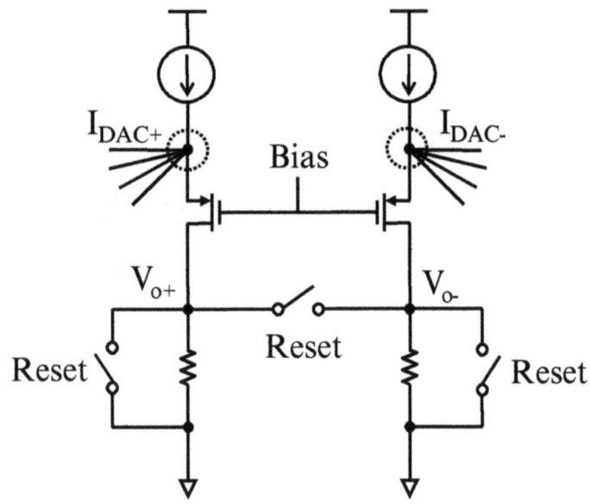

Fig. 4.33 Current-steering differential RZ DAC

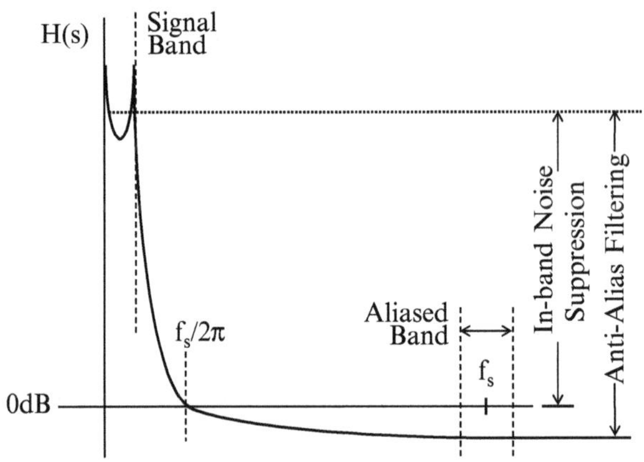

Fig. 4.34 In-band gain of the CT loop filter and its anti-aliasing

4.5.3 Self-Trimming Time Constants

SC DSMs have been staples at low frequencies for voice and audio applications, but CT DSMs are gaining momentum and widely used in wireless receivers. CT loop filters perform both anti-aliasing and noise-shaping functions as shown in the example of a third-order loop filter in Fig. 4.34. In addition, the STF as sketched in Fig. 4.35 can be also shaped to suppress neighboring blockers further using zeros at blocker channels. Loop filters of the oversampling DSM can be realized using either CT or SC integrators as shown in Fig. 4.36.

The unity-gain bandwidth of the DSM loop filter is set to be $f_s/2\pi$, which is about 2π times lower than the sampling clock frequency. That is, its time constant is simply $1/f_s$. In the CT integrator, the time-constant value of the integrator becomes $RC = 1/f_s$. In the SC integrator, the loop bandwidth is set by sizing the sampling capacitor C_s to be the same as the integrating capacitor C_i. Then these two integrators implement an identical loop filter. The time constant C_i/f_sC_s of the SC

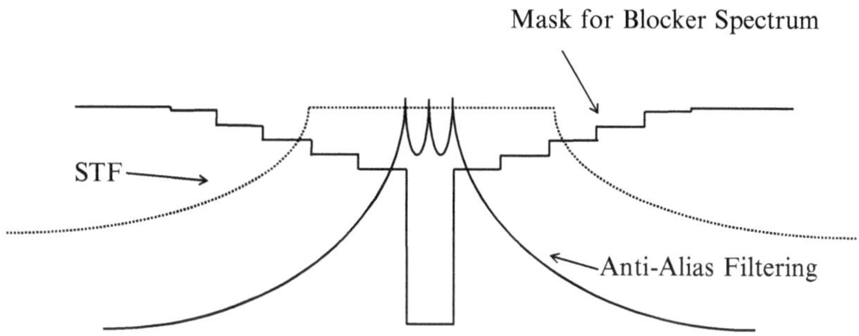

Fig. 4.35 Blocker mask with anti-aliasing and STF

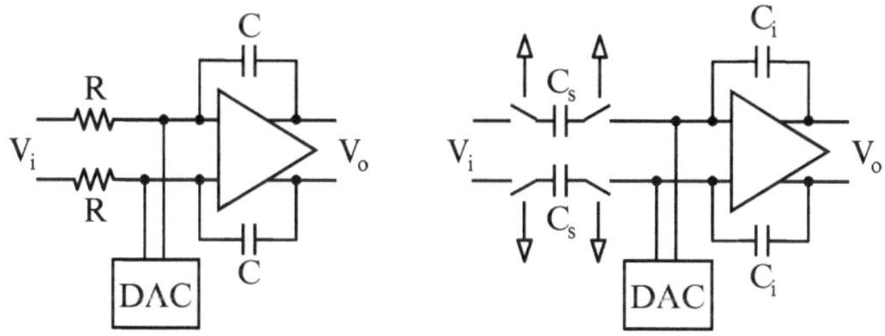

Fig. 4.36 CT and SC Integrators

integrator is set by the accurate capacitor ratio and the clock frequency of crystal precision.

However, since the time constant of the CT integrator is set by the absolute values of R and C, the unity loop-gain frequency of the CT loop filter is very sensitive to process variations. The DSM performance is directly affected by the loop gain and bandwidth. If the time constant is too short, the loop bandwidth gets wider, and the loop becomes unstable. On the other hand, if too long, the in-band noise is not suppressed as much as designed. Therefore, it is a common practice to overdesign the loop bandwidth to be wider than the desired. The process sensitivity of the loop filter can be greatly reduced by self-trimming uncertain time constants. It is the most common classical analog problem of trimming or calibrating time constants of CT filters. The effect of the loop filter variations over process is illustrated in Fig. 4.37.

The most common way to trim the time constant is to have a replica of the integrator, and use it as a master. The time constant can be measured directly, or an oscillator is locked to the time constant so that actual slave cells can be controlled using the same voltage or digital settings. Such master/slave approaches are limited by the matching accuracy of the time constant. A better way is to inject a test tone at a zero frequency or out of bandwidth, and detect the tone magnitude at the output to detect the time constant.

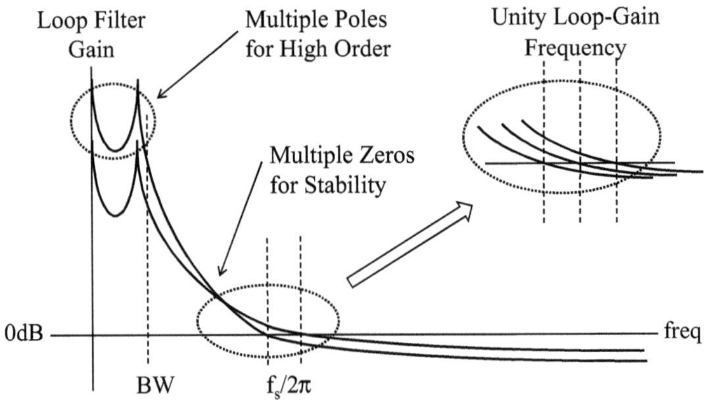

Fig. 4.37 Variations of loop filter over process

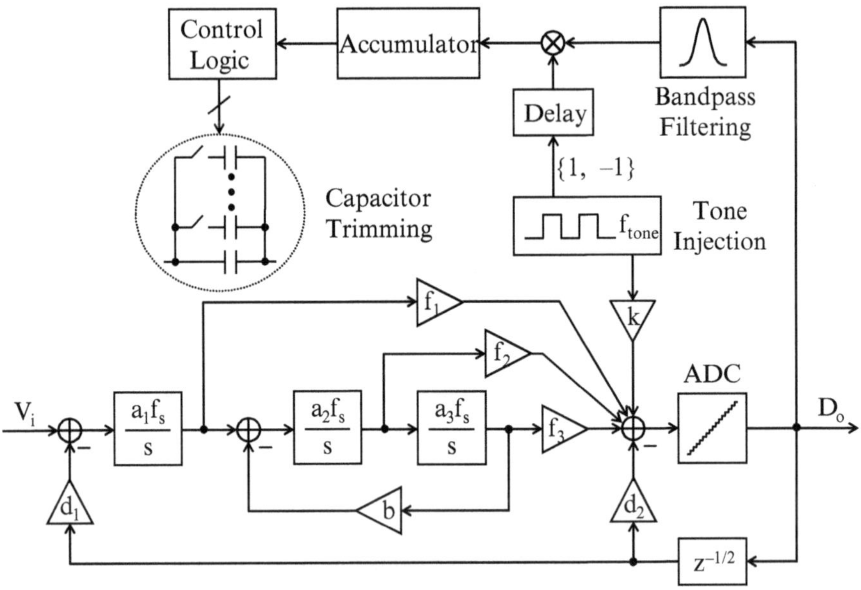

Fig. 4.38 Time-constant self-trimming by tone injection at zero frequency

A small binary pulse tone f_{tone} can be injected into the comparator input for self-trimming as shown in the example of a third-order modulator with a resonator pole at the signal band edge of Fig. 4.38 [29]. The tone can be of any magnitude unless the headroom is limited. It is quantized together with the quantization error, and suppressed by the same NTF. The self-trimming loop is to align the tone frequency with the actual zero frequency f_{zero} by trimming the integrator *RC* time constant. The single tone changes its phase by 180° at the center frequency of the bandpass filter. Therefore, multiplying the bandpass filtered tone by the

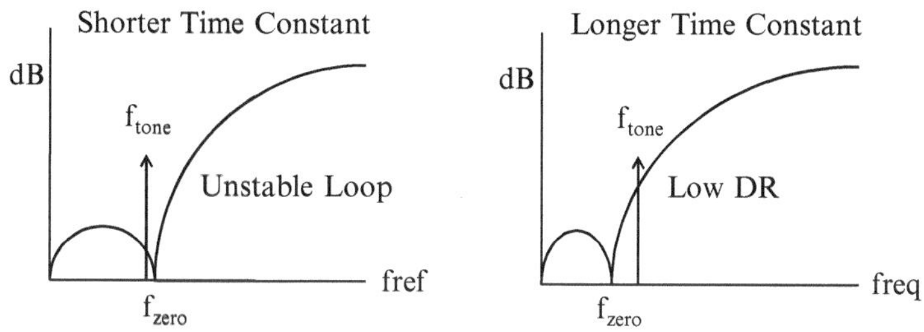

Fig. 4.39 Injected tone and zero frequency

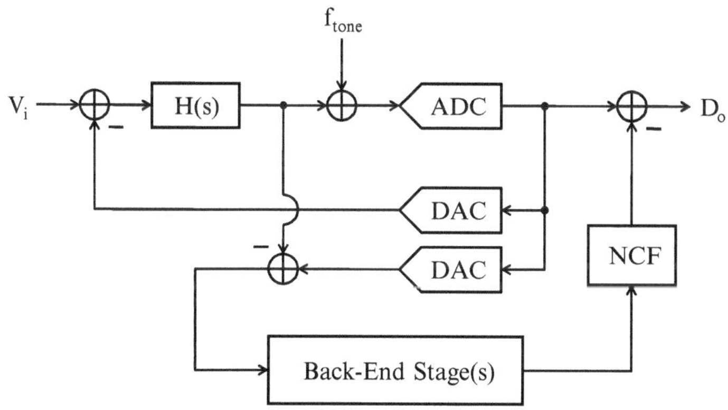

Fig. 4.40 Tone injection for self-trimming in cascaded modulator

delay-matched injected tone becomes the error power whose polarity changes at the bandpass center frequency. The relationship between f_{tone} and f_{zero} are illustrated in Fig. 4.39.

If $f_{tone} < f_{zero}$, the time constant is shorter than the desired, and the loop can become unstable. Otherwise, the loop filter gain is lower, and the dynamic range can be reduced. The effectiveness is only limited by the step size of the time-constant adjustment. If any fixed input stays exactly on the zero frequency, the tone rejection can be incomplete. However, in modern spread-spectrum systems, most input spectral densities are random and broadbanded, and it is very unlikely to have a fixed constant input tone all the time at the zero frequency in particular.

The same self-trimming by tone injection can be applied to cascaded modulators as shown in Fig. 4.40 [30]. However, the injected tone doesn't need to be the in-band zero frequency. It can be of any out-of-band frequency. The error detection mechanism is the same. In the cascaded modulator, the first-stage quantization error becomes the residue, and quantized again by the second and the following stages. Since the tone is quantized by the first and later stages, the working principle is based on the matching of the two gain paths. That is, the tone should disappear from the output if two path time constants are matched. If the first-stage quantization noise is cancelled, the injected tone is also cancelled. So the self-trimming is to detect the existence of the residual tone at the output, and adjust the time constant of the first-stage loop filter.

References

1. B. Song, M. Tompsett, K. Lakshmikumar, A 12-Bit 1-Msample/s capacitor error averaging pipelined A/D converter. IEEE J. Solid State Circuits **SC-23**, 1324–1333 (1988)
2. B. Song, S. Lee, M. Tompsett, A 10b 15MHz CMOS recycling two-step A/D converter. IEEE J. Solid State Circuits **SC-25**, 1328–1338 (1990)
3. L. Brooks, H. Lee, A zero-crossing-based 8-bit 200MS/s pipelined ADC. IEEE J. Solid State Circuits **SC-42**, 2677–2687 (2007)
4. F. van der Goes et al., A 1.5mW 68dB SNDR 80MS/s 2x interleaved SAR-assisted pipelined ADC in 28nm CMOS, *ISSCC Dig. Tech. Papers* (Feb 2014), pp. 200–201
5. D. Schinkel, E. Mensink, E. Klumperink, E. van Tuijl, B. Nauta, A double-tail latch-type voltage sense amplifier with 18ps setup + hold time, *ISSCC Dig. Tech. Papers* (Feb 2007), pp. 314–315
6. J. McCreary, P. Gray, All-MOS charge redistribution analog-to-digital conversion techniques, I. IEEE J. Solid State Circuits **SC-10**, 371–379 (1975)
7. J. Candy, A use of double integration in sigma-delta modulation, IEEE Trans. Comm. **COM-33**, 249–258 (1985)
8. K. Cao, S. Nadeem, W. Lee, C. Sodini, A higher order topology for interpolative modulators for oversampling A/D converters. IEEE Trans. Circuits Syst. **37**, 309–318 (1990)
9. T. Hayashi, Y. Inabe, K. Uchimura, T. Kimura, A multi-stage delta-sigma modulator without double integrator loop, *ISSCC Dig. Tech. Papers* (Feb 1986), pp. 182–183
10. M. Hesener, T. Eichler, A. Hanneberg, D. Herbison1, F. Kuttner, H. Wenske, A 14b 40MS/s redundant SAR ADC with 480MHz clock in 0.13μm CMOS, *IEEE Dig. Tech. Papers* (Feb 2007), pp. 248–249
11. M. Kramer, E. Janssen, K. Doris, B. Murman, A 14b 35MS/s SAR ADC achieving 75dB SNDR and 99dB SFDR with loop-embedded input buffer in 40nm CMOS, *ISSCC Dig. Tech. Papers* (Feb 2015), pp. 284–285
12. M. Park, M. Perrott, A 0.13μm CMOS 78dB SNDR 87mW 20MHz BW CT ΔΣ ADC with VCO-based integrator and quantizer, *ISSCC Dig. Tech. Papers* (Feb 2009), pp. 170–171
13. H. Schmid, *Electronic Analog/Digital Conversion Techniques* (Van Nostrand, New York, 1970), pp. 323–325
14. S. Lewis, P. Gray, A pipelined 5-Msample/s 9-bit analog-to-digital converter. IEEE J. Solid State Circuits **SC-22**, 954–961 (1987)
15. AD7960, LTC2389-18, TI-ADS8881, Data Sheets
16. C. Hurrell, C. Lyden, D. Laing, D. Hummerston, M. Vickery, An 18b 12.5MHz ADC with 93dB SNR, *ISSCC Dig. Tech. Papers* (Feb 2010), pp. 378–379
17. V. Quiquempoix, P. Deval, A. Barreto, G. Bellini, J. Márkus, J. Silva, Member, G. Temes, A low-power 22-bit incremental ADC, IEEE J. Solid State Circuits **41**, 1562–1572 (2006)
18. A. Agah, K. Vleugels, P. Griffin, M. Ronaghi, J. Plummer, B. Wooley, A high-resolution low-power incremental ΣΔ ADC with extended range for biosensor arrays. IEEE J. Solid State Circuits **45**, 1099–1110 (2010)
19. C. Lee, M. Flynn, A 14 b 23 MS/s 48 mW resetting ADC. IEEE Trans. Circuits Syst I **58**, 1167–1177 (2011)
20. T. Katayama, K. Koyama, An 18b, 1.85MS/s incremental ΔΣ ADC with input S/H, *AKM Microdevices* (Aug 2015)

21. K. Nam, S. Lee, D. Su, B. Wooley, A low-voltage low-power sigma-delta modulator for broadband analog-to-digital conversion. IEEE J. Solid State Circuits **SC-40**, 1855–1864 (2005)
22. J. Domogalla, Combination of analog to digital converter and error correction circuit, U.S. Patent 4 451 821 (May 1984)
23. H. Lee, D. Hodges, P. Gray, A self-calibrating 15 bit CMOS A/D converter. IEEE J. Solid State Circuits **SC-19**, 813–819 (1984)
24. S. Lee, B. Song, Digital-domain calibration techniques for multi-step A/D converters. IEEE J. Solid State Circuits **SC-27**, 1679–1688 (1992)
25. T. Shu, B. Song, K. Bacrania, A 13-b 10-Msample/sec ADC digitally calibrated with real-time oversampling calibrator. IEEE J. Solid State Circuits **SC-30**, 443–452 (1995)
26. A. Bugeja, B. Song, A self-trimming 14b 100MS/s CMOS DAC. IEEE J. Solid State Circuits **SC-35**, 1841–1852 (2000)
27. M.J. Choe, B. Song, K. Bacrania, A 13b 40MS/s CMOS pipelined folding ADC with background offset trimming. IEEE J. Solid State Circuits **SC-35**, 1781–1790 (2000)
28. S. Ryu, S. Ray, B. Song, G. Cho, K. Bacrania, A 14b-linear capacitor self-trimming pipelined ADC. IEEE J. Solid State Circuits **SC-39**, 2046–2051 (2004)
29. Y. Shu, B. Song, K. Bacrania, A 65nm CMOS CT $\Delta\Sigma$ modulator with 81dB DR and 8MHz BW auto-tuned by pulse injection, *ISSCC Dig. Tech. Papers* (Feb 2008), pp. 500–501
30. Y. Shu, J. Kamiishi, K. Tomioka, K. Hamashita, B. Song, LMS-based noise leakage calibration of cascaded continuous-time $\Delta\Sigma$ modulators. IEEE J. Solid State Circuits **SC-45**, 368–379 (2010)

Chapter 5
Switched-Capacitor Circuits

The elusive goal of the lossless charge transfer from one circuit node to another has been the main focus of analog sampled-data processing starting from the bucket brigade and CCD to switched-capacitor circuits. Opamp with capacitive feedback is the most accurate analog component that can sample and amplify signal with highest accuracy. Their voltage transfer accuracy is solely dependent upon the DC gain and nonlinearity of the opamp. The lossless voltage transfer regardless of the opamp gain and nonlinearity error is achievable by eliminating it right from its source in the analog domain. When applied to the pipelined ADC, the linearity performance can be enhanced by adaptively cancelling them based on the global zero-forcing LMS feedback. Two featured circuit concepts can be incorporated to implement error-free switched-capacitor amplifiers.

5.1 Analog Sampled-Data Processing

Sampled-data signal processing is an analog equivalence to digital signal processing. Instead of quantizing the analog sampled DC signal for digital processing, it keeps it in the analog domain for discrete-time analog signal processing. It requires three basic circuit components to perform sample and hold, amplify, and integrate. For integrated circuits, the only analog energy storage to hold sampled signal is the capacitor. Two components of a capacitor and MOS switch made an analog delay element called bucket brigade, which also performed the primitive add and subtract function. The sampled-data analog processing has been evolved quickly for practical uses like image processing as the charge-coupled device was introduced. However, as CMOS analog technology advanced, switched-capacitor circuits using feedback opamps have become the de facto standard for analog sampled-data signal processing.

In switched-capacitor circuits, the sample/hold is just to freeze analog waveform in time, and the amplifier is to perform all basic functions to add, subtract, and

B.-S. Song, *System-level Techniques for Analog Performance Enhancement*,
DOI 10.1007/978-3-319-27921-3_5

amplify analog sampled signals called sampled data like digital numbers. The integrator can be implemented both in continuous time or discrete time. Integrator and integrate/dump are the most common components in digital processing. In analog, they are used only in feedback configuration, but seldom used as stand-alone open-loop devices since they are DC unstable. Therefore, the absolute gain of the integrator is not critical in feedback such as $\Delta\Sigma$ modulators and error amplifiers for control. In switched-capacitor amplifiers, both the linearity and absolute gain matter as most applications are open-loop cases. However, the absolute gain of the sample/hold matters little though they are used in open-loop condition, but its linearity often limits the system performance.

Switched-capacitor amplifiers perform the basic function of transferring a sampled analog signal from one circuit node to another. For that, the charge sampled on the input sampling capacitor should be completely transferred to the feedback capacitor so that the signal gain can be accurately set by the capacitor ratio only. The difficulty lies in how to design them to achieve both high accuracy and high speed simultaneously for settling. A viable solution that enables the high-resolution analog processing at premium speeds is to alleviate the gain and bandwidth requirement of the opamp by eliminating the settling error at the opamp input summing node. In switched-capacitor circuits, opamp works as an error amplifier, and forces the capacitive opamp input summing node to settle to the virtual ground voltage. However, due to the finite gain and bandwidth of the opamp, the summing node doesn't settle to the ideal ground voltage, and it also takes time for the summing node to settle. The former makes the gain and nonlinearity error in the settled output, and the latter limits the sampling rate.

Most common solutions try to shorten the initial nonlinear slewing time using class AB opamps or nonlinear switching amplifiers. Although the slew-rate enhancement shortens the initial settling time for large inputs, the bandwidth of the amplifier still affects the linear settling time. Since the slewing time is usually much shorter than the linear settling time in high-resolution cases, the high bias current is required for broadbanding, and opamp designs with high bias currents are inevitably subject to the gain and bandwidth trade-off. For this reason, alternative architectures such as successive approximation, time-domain, or comparator-based circuits have been sought after to entirely avoid using opamps mostly for low-power applications.

5.2 Opamp-Induced Gain Error

Switched-capacitor circuits are two-phase DC circuits driven by two nonoverlapping clocks. One phase is the sampling phase, and the other is usually the amplifying or integrating phase. At the beginning of each phase, they start from new initial conditions on capacitors. After the transient period, all nodes settle to final DC values. As a result, the final voltage at the output node is sampled by the next stage. Once fully settled, the output of the switched-capacitor amplifier is a DC

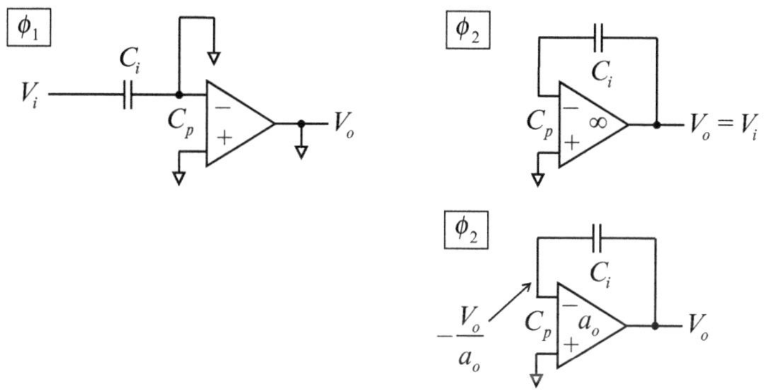

Fig. 5.1 Switched-capacitor sample/hold in two clock phases

voltage, and its accuracy is limited by the capacitor mismatch and the opamp gain and nonlinearity.

Figure 5.1 illustrates the standard switched-capacitor sample/hold circuits in two nonoverlapping clock phases, ϕ_1 and ϕ_2. Note that switches are not drawn in the figure for simplicity. During the sampling phase ϕ_1, the input V_i is sampled on the input capacitor C_i while the opamp input and output nodes are reset to ground. Then the charge Q sampled on C_i is $Q = C_\mathrm{i}V_\mathrm{i}$. During the amplification phase ϕ_2 as shown on the right side, the sampling capacitor is flipped and connected to the output. As shown on the top right, the held output voltage V_o after the output settles will be $V_\mathrm{o} = V_\mathrm{i}$ if the opamp is ideal with infinite gain. However, if the opamp has finite gain a_o as shown on the bottom right, the input summing node of the opamp is lowered to $-V_\mathrm{o}/a_\mathrm{o}$. If the total parasitic capacitance at the summing node is C_p, the charge conservation rule gives

$$Q = C_\mathrm{i}V_\mathrm{i} = C_\mathrm{i}\left(V_\mathrm{o} + \frac{V_\mathrm{o}}{a_\mathrm{o}}\right) + C_\mathrm{p}\frac{V_\mathrm{o}}{a_\mathrm{o}}. \tag{5.1}$$

Solving for V_o, the output includes the opamp gain and nonlinearity error.

$$V_\mathrm{o} = \frac{V_\mathrm{i}}{1 + \frac{C_\mathrm{i}+C_\mathrm{p}}{a_\mathrm{o}C_\mathrm{i}}} = \frac{V_\mathrm{i}}{1 + \frac{1}{a_\mathrm{o}f}} = \frac{V_\mathrm{o}|_\mathrm{Ideal}}{1 + \frac{1}{a_\mathrm{o}f}}, \tag{5.2}$$

where the feedback factor is defined as

$$f = \frac{C_\mathrm{i}}{C_\mathrm{i} + C_\mathrm{p}}, \tag{5.3}$$

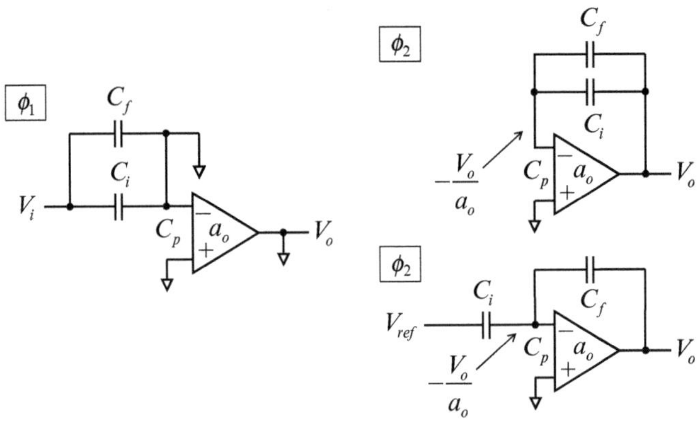

Fig. 5.2 Switched-capacitor amplifier in two clock phases

and

$$V_o|_{\text{Ideal}} = V_i = V_o\left(1 + \frac{1}{a_o f}\right) = V_o - \left(-\frac{V_o}{a_o f}\right). \tag{5.4}$$

That is, the summing node error produces the gain and nonlinearity error of $-V_o/a_o f$ at the output. In principle, a smaller replica of the gain and nonlinearity error appears at the summing node after attenuated by f in the feedback path.

The same figure is repeated in Fig. 5.2 for the switched-capacitor amplifier [1]. On the right side, the sample/hold and amplifier configurations are shown at the top and bottom, respectively. The same equations as (5.1) through (5.4) hold true also for the sample/hold case if the capacitor C_i is replaced as $C_i + C_f$. Similarly, the equations for the amplifier shown on the right bottom can be derived as follows. The charge conservation in two clock phases gives

$$(C_i + C_f)V_i = (C_i + C_p)\frac{V_o}{a_o} + C_f\left(V_o + \frac{V_o}{a_o}\right). \tag{5.5}$$

Solving for V_o yields the residue output as follows.

$$V_o = \frac{C_i + C_f}{C_f} \times \frac{V_i}{\left(1 + \frac{1}{a_o f}\right)} = \frac{V_o|_{\text{Ideal}}}{\left(1 + \frac{1}{a_o f}\right)}, \tag{5.6}$$

where the feedback factor f is defined as

$$f = \frac{C_f}{C_i + C_f + C_p}, \tag{5.7}$$

and

$$V_o|_{\mathrm{Ideal}} = \frac{C_i + C_f}{C_f} \times V_i = V_o\left(1 + \frac{1}{a_o f}\right) = V_o - \left(-\frac{V_o}{a_o f}\right). \tag{5.8}$$

Note that (5.8) is identical to (5.4) except for the ideal error-free output value $V_o|_{\mathrm{Ideal}}$, which tells the difference between the switched-capacitor sample/hold and amplifier. The amplifier gain is set accurately by the capacitor ratio, which is the very advantage of the discrete-time switched-capacitor circuit over other continuous-time amplifiers with resistive feedback.

The effect of the opamp finite gain and nonlinearity can be better understood by defining the error voltage occurring at the input summing node of the opamp. When the amplifier settles, the opamp input node undergoes the same error of $-V_o/a_o$ as the output in a smaller scale since the output error is attenuated by the feedback factor f. The actual output suffers the signal loss of $-V_o/a_o f$ that is inversely proportional to the loop gain. That is, the opamp-induced gain error is the output attenuated by the feedback loop gain. The gain factor of $1/f$ is to compensate for the attenuation loss in the feedback path. This relation gives a clue to how to eliminate the total opamp gain and nonlinearity error in the analog domain.

The feedback factor f is defined as the attenuation of the feedback path. It includes the opamp input capacitance including the Miller effect plus the stray capacitance of the capacitor top plates and the parasitic capacitances of the turned-off initializing switch as shown in Fig. 5.3.

The opamp input Miller capacitance is dominant unless the opamp input is buffered using a source follower or cancelled using a negative Miller capacitance. The cancellation of the Miller effect is commonly done by inserting two normally off-transistor capacitors of C_{gd} between one differential input and the drain of the other input transistor. The replica capacitance of C_{gd} can be made of a half-sized input transistor with its source and drain tied together. If it is turned off, the channel overlap capacitance of the half-sized device is the same as that of the differential input device.

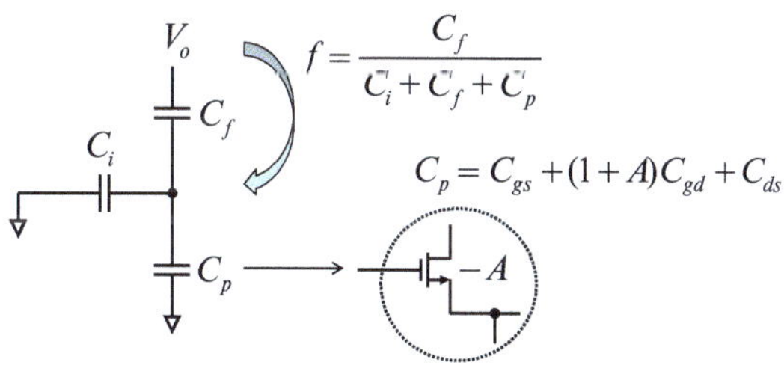

Fig. 5.3 Parasitic capacitance at the opamp input summing node

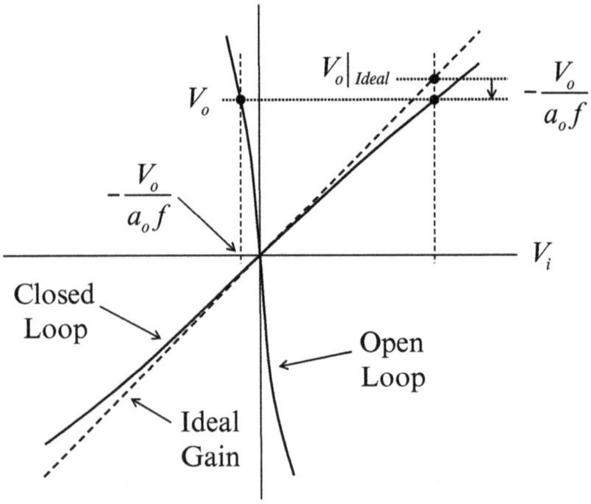

Fig. 5.4 Open- and closed-loop transfer characteristics

Figure 5.4 illustrates the gain-related error graphically using both open- and closed-loop characteristics. Assume that the amplifier settles to the output V_o in two transfer curves. In the open-loop curve of the loop gain $a_o f$, the input summing node voltage shifts down by $-V_o/a_o f$ as shown on the x-axis, and in the closed-loop, the output voltage also shifts down by the same amount of $-V_o/f$. Thus the ideal $V_o|_{\mathrm{Ideal}}$ can be obtained regardless of the opamp gain and nonlinearity error by adding this error back to the output. Any switched-capacitor feedback circuit with a DC feedback loop gain of $a_o f$ makes an error of $-V_o/a_o f$, which is proportional to the output but inversely proportional to the loop gain. If this error is precisely reproduced and added back to the output, the lossless signal transfer regardless of the opamp gain and nonlinearity is accomplished.

5.3 Accurate Interstage Residue Transfer

The switched-capacitor amplifier shown in Fig. 5.2 is the basic functional block to implement all sampled-data circuits such as precision instrument amplifier and capacitor-array multiplying DAC (MDAC). The most prominent use of the latter is an interstage residue amplifier in the pipelined ADC. The residue in the pipelined ADC operation is the unquantized portion of the signal left for the back-end fine quantization stages to quantize. Thus the function of the residue amplifier is to accurately pipeline a residue to the subsequent stage. The static linearity of the pipelined ADC is mainly affected by the accuracy of the DC voltage transferred from the first-stage output to the next-stage sampling capacitor. The MDAC-based switched-capacitor amplifier is perfect for this use as it performs the function of transferring accurate residues from stage to stage.

The function of MDAC is to amplify the sampled input with an exact gain of binary number, and subtract the binary multiples of the reference voltage. Its

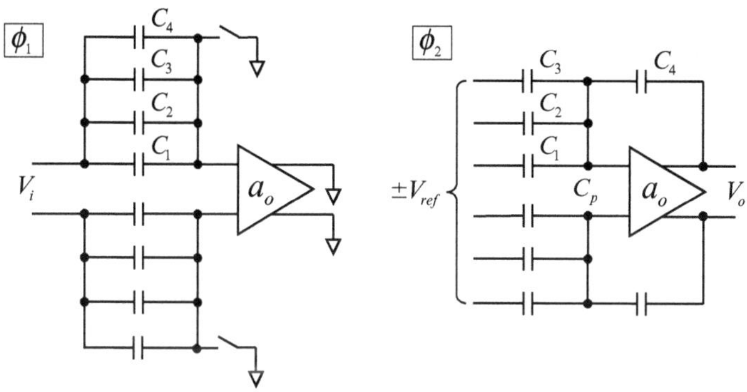

Fig. 5.5 Differential 2b switched-capacitor MDAC in two clock phases

switched-capacitor implementation is subject to two error sources that result from the capacitor mismatch and the gain and nonlinearity of the opamp. Although the capacitor matching accuracy has been constantly improved over the years, it has been difficult to exclude the opamp gain and nonlinearity issue from the high-resolution pipelined ADC design. The fact is that the situation got worse as the technology was scaled down to the nanometer range and the supply voltage headed towards below 1 V. At low voltages, opamp operation is severely handicapped with low gain and limited signal swing, and even the trend of opamp-free analog designs has emerged and created completely unconventional designs of switching nonlinear analog circuits.

A differential 2b switched-capacitor MDAC is explained in Fig. 5.5, which performs as a residue amplifier and a DAC for pipelined stages. It is the same switched-capacitor amplifier shown in Fig. 5.2 in its operation. During one-clock phase, the input V_i is sampled on all bottom plates of a capacitor-array DAC made of four unit capacitors by initializing the common top plate tied to the input summing node to ground. At this time, the opamp output is also grounded to prevent the DC wandering at the output and to reset the initial condition of the output before transient. During the other clock phase, with one capacitor C_4 connected to the output, the bottom plates of the other three capacitors are driven by the reference voltages depending on the coarse sub-ADC digital output. If the opamp is ideal, the 2b MDAC output settles to yield

$$V_o = 2^2 \times (V_i - DV_i), \tag{5.9}$$

where DV_i is the DAC output representing the 2b digitized input [2]. This 2b residue is the unquantized portion of the signal that the later stages resolve further for finer resolution. In general for N-b DACs, the total number of the sampling capacitors in the array is 2^N for two-level DACs ($\pm V_{ref}$), and 2^{N-1} for tri-level DACs ($\pm V_{ref}$, 0).

Now considering the DC gain a_o and the parasitic at the input summing C_p, equations from (5.5) to (5.8) are true if the ideal output is modified as follows.

$$V_o|_{\text{Ideal}} = \frac{C_1 + C_2 + C_3 + C_4}{C_4} \times V_i = V_o\left(1 + \frac{1}{a_o f}\right) = V_o - \left(-\frac{V_o}{a_o f}\right), \quad (5.10)$$

where the feedback factor f is defined as

$$f = \frac{C_4}{C_1 + C_2 + C_3 + C_4 + C_p}. \quad (5.11)$$

In (5.11), the only source of nonlinearity that affects f is the summing-node parasitic capacitance C_p. However, the nonlinearity in the feedback factor f is negligible since the summing node of the opamp experiences no significant voltage change due to the high feedback loop gain, and the differential circuitry cancels the even-order nonlinearity in f to the first order. Thus it can be safely assumed that the feedback factor f stays constant. So does the closed-loop gain of $1/f$. Note that in (5.10), the ideal residue gain is set by the capacitor ratio if the opamp gain-related error is eliminated.

5.4 History of Opamp-Induced Gain Error Cancellation

Since early 1980s, analog designs have evolved around high-resolution techniques for switched-capacitor circuits, and the capacitor mismatch and opamp finite gain and nonlinearity have been overcome indirectly using either sophisticated digital processing or oversampling techniques such as ΔΣ modulation. In 1980s when the technology feature sizes were 1.2–5 μm, the lithography was limited, and the unit capacitor was sized to be about 25 μm by 25 μm to barely get 8–10b resolution. The pipelined ADC was clocked at 20 MS/s consuming about 250 mW at 3.3 V. Technology scaling in 1990s made submicron feature sizes of 0.25–0.8 μm with a unit capacitance of 12 μm by 12 μm, and enhanced the ADC performance up to 10–12b at 20–50 MS/s while consuming only 20–150 mW at 1.8 V. Up to this point, the finite opamp gain was not a significant issue though the capacitor matching accuracy has improved significantly. After 2000, the technology feature size was scaled further down to the nanometer range of 65–24 nm with a shrinking unit capacitance of only 5 μm by 5 μm, and enabled 12–14b ADCs with bare capacitor matching while clocking at 50–250 MS/s and consuming only 10–250 mW with 1–1.2 V supply.

Device scaling played a major role in achieving high resolution and high speed with small area and low power. Fine lithography offered unprecedented speed and matching advantages, but taxed analog designs heavily with low voltage and gain constraints. Alternative solutions have also been suggested to digitally calibrate opamp gain and nonlinearity, to remove the low voltage and gain constraints by self-trimming, or to use opamp-free switching-based analog system architectures such as SAR, time-domain, and comparator-based ADCs. However, the opamp-induced gain error still stays as the most critical factor in switched-capacitor circuit designs. There was no known analog method until recently that removes the error resulting from finite DC gain and nonlinearity.

Both analog and digital methods have been investigated to remove the gain error of the opamp at the ADC system level [3–5]. In the digital approach, the transfer function of the residue amplifier is modeled to be weakly nonlinear with a dominant third harmonic [6–8]. DC parameters such as the offset, the linear gain error, and the third harmonic are individually measured and cancelled later. However, the nonlinearity error characterized only by the third harmonic coefficient requires sophisticated numerical processing to estimate the actual error that is proportional to the cubic term of the output. That is, the nonlinearity errors should be distributed over the full output range either using look-up tables or through numerical calculations. Furthermore, even measuring only one third harmonic coefficient is a daunting task by itself. Either code density or pseudo-random dithering is used, but it requires a rare condition that the input signal is well-defined and busy.

In the analog approach, the opamp summing-node error can be treated as a separate signal, and the feedforwarded error can be subtracted from the residue output. This two-path analog system works if the gain of the auxiliary error path is accurately matched to the gain of the main signal path. Although two-path gains are difficult to match, the auxiliary gain path can be implemented using another switched-capacitor amplifier with an estimated gain. Even coarse matching with a ± 10 % accuracy will provide about 20 dB gain improvement. However, the switched-capacitor circuit is noisy since it suffers extra sampling-related noises such as kT/C and sampled wideband noise. Instead the digital approach can be employed to measure the opamp summing-node error accurately using another slow but fine-resolution ADC, and the summing-node error is cancelled either by raising the actual gain of the opamp with positive feedback or by calibrating it digitally. An analog feedback approach offers a closed-loop solution to the problem.

5.5 Nonlinearity-Cancelled Bottom-Plate Sampling

In the standard version of the bottom-plate sampling, the signal is sampled on the bottom plates of all sampling capacitors, and the charge injection and clock feed-through errors at the common top plate are kept constant and independent of the input since they are initialized first. When sampling the AC input, the nonlinearity of the bottom-plate switch on-resistance affects the linearity of the sampled voltage as the threshold voltages of the switches are modulated by the AC input. The effect of the switch nonlinearity is alleviated if the on-resistance can be made smaller or more linear either using large switches or boosting the clock with a constant overdrive voltage. However, even the standard sample/hold is not immune to the opamp-induced gain error as the next-stage samples the very output of the opamp.

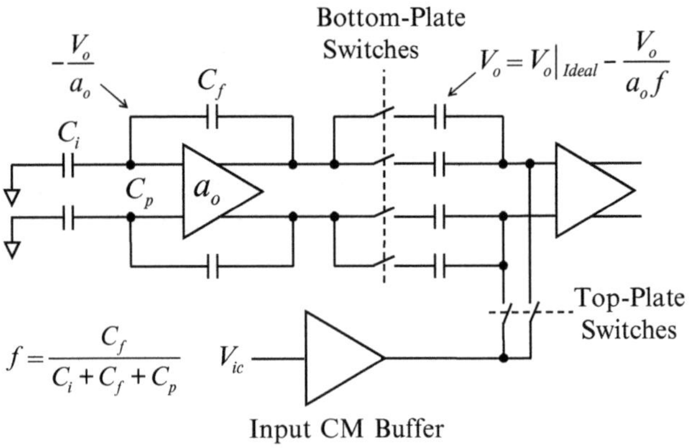

Fig. 5.6 Standard bottom-plate sampling

Self-trimming is an electronic version of the physical trimming process that nulls the opamp gain and nonlinearity error. First, the MDAC output is sampled after its opamp-induced error is cancelled by its estimate, and some residual error is left in the digital output. By detecting the residual error from the digitized output, the gain of the error path is adjusted using a zero-forcing LMS feedback algorithm [9]. A simple differential pair can be used as an analog programmable gain element. If the top plate of the next-stage sampling capacitor is initialized with the error of the previous stage, the sampled residue is completely free from the previous-stage opamp error. That is, the total error of the residue amplifier can be precisely cancelled only by setting the gain of the error path correctly.

Figure 5.6 illustrates the standard bottom-plate sampling scheme used in most high-resolution switched-capacitor residue amplifiers [1, 2]. The top-plate switches that initialize the input CM voltage are turned off slightly earlier than the bottom-plate switches to make the switch charge injection and clock feedthrough voltage constant and independent of the input [10]. It achieves the highest level of linearity in sampling, and is known to be the most accurate sample/hold circuit in CMOS analog designs. Since the next-stage samples the differential output of the previous stage relative to the constant ground, the voltage sampled by the next stage is obtained from (5.8).

$$V_\mathrm{o} = \frac{C_\mathrm{i} + C_\mathrm{f}}{C_\mathrm{f}} V_\mathrm{i} - \frac{V_\mathrm{o}}{a_\mathrm{o} f} = V_\mathrm{o}|_\mathrm{Ideal} - \frac{V_\mathrm{o}}{a_\mathrm{o} f}. \tag{5.12}$$

That is, the finite gain and nonlinearity error inversely proportional to the loop gain is transferred to the next stage in this standard bottom-plate sampling.

Equation (5.12) shows that the signal transfer loss can be reconstructed if the opamp input summing-node error is amplified by the linear gain of $1/f$. Note also that it holds true without regard to whether the open-loop DC gain characteristic of the opamp is linear or not, and the total error is $-V_\mathrm{o}/a_\mathrm{o} f$. That is, if the exact transfer

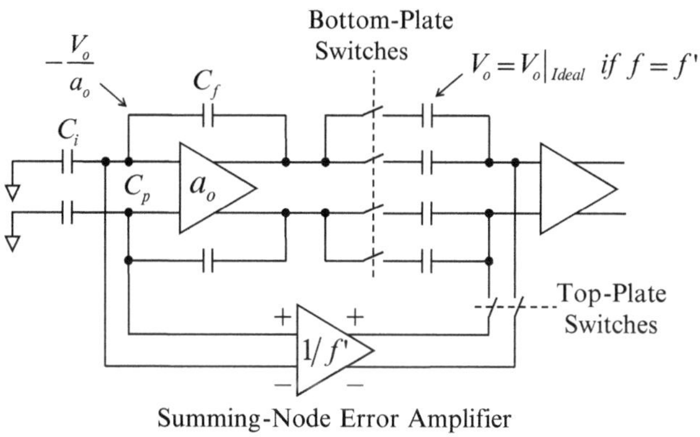

Fig. 5.7 Lossless bottom-plate sampling

loss is replicated and added back to the output, the error-free ideal residue output can be restored. However, there is no way of knowing or measuring the exact gain of $1/f$. The only viable option is to estimate the gain error first and update it employing the LMS servo-feedback algorithm that can force the residual error to converge to zero.

Lossless bottom-plate sampling that is free from the opamp-induced error is shown in Fig. 5.7. In this arrangement, the only change made is an internal CM buffer that usually buffers a constant voltage and initializes the top plates is replaced by the summing-node error amplifier. Thus the total error occurring at the summing node is fed forward with a gain of $1/f'$ so that the nonlinearity-free residue can be directly sampled on the capacitance of the next stage. The estimated linear gain of $1/f'$ should closely match with the main path gain of $1/f$. Then from (5.12), the actual voltage to be sampled on the next-stage sampling capacitor becomes

$$V_{\text{sampled}} = \left(\frac{C_{\text{i}} + C_{\text{f}}}{C_{\text{f}}} V_{\text{i}} - \frac{V_{\text{o}}}{a_{\text{o}} f}\right) - \left(-\frac{V_{\text{o}}}{a_{\text{o}} f'}\right)$$
$$= V_{\text{o}}\big|_{\text{Ideal}} - \frac{V_{\text{o}}}{a_{\text{o}}} \times \left(\frac{1}{f} - \frac{1}{f'}\right) = V_{\text{o}}\big|_{\text{Ideal}}, \quad \text{if } \frac{1}{f'} = \frac{1}{f}. \tag{5.13}$$

From (5.13), an important conclusion can be drawn. The original goal to cancel the gain and nonlinearity error of the opamp has now turned into a new goal to match the two linear path gains of $1/f$ and $1/f'$. If they are perfectly matched, the opamp-induced gain error disappears as the gain of the opamp becomes infinite in effect. That is, the effective gain can be defined as

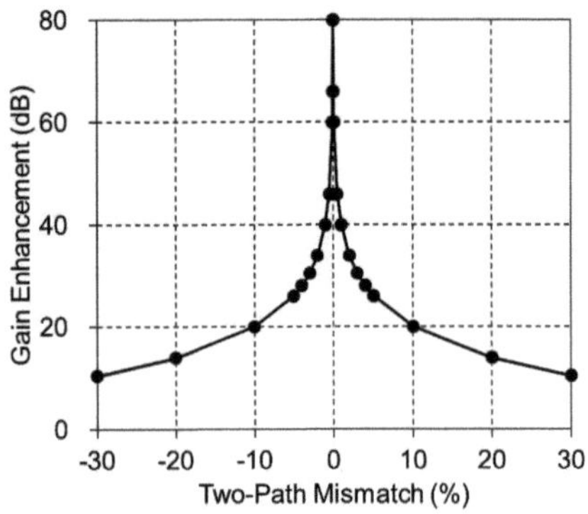

Fig. 5.8 Enhancement factor for the effective gain

$$a_{\text{eff}} = \left| \frac{a_o}{f(1/f - 1/f')} \right| = \left| \frac{a_o f'}{(f' - f)} \right| \approx a_o \times \frac{f}{\Delta f}. \tag{5.14}$$

As expected, it depends on the path gain mismatch of $\Delta f/f$, and therefore the effect of the gain enhancement peaks when they are matched as shown in Fig. 5.8.

5.6 LMS Adaptation for Gain and Nonlinearity Error

Therefore, it all boils down to how to match two-path gains precisely. They can be adaptively matched by self-trimming the summing-node amplifier gain $1/f'$. The concept of linearizing the transfer function is illustrated in Fig. 5.9 [9, 11].

The sign–sign LMS adaptation of the opamp gain and nonlinearity error requires the error polarity detection. The same error goes through two different gain paths, and the split errors are subtracted when sampled on the sampling capacitor of the next stage. The physical gain of the summing-node amplifier can be updated using an up/down counter by comparing the measured voltage DV_{ref} to the ideal V_{ref} as shown in Fig. 5.10.

As the gain of the summing-node amplifier increases, the nonlinearity of the transfer function is cancelled gradually. From (5.13), if $1/f' = 1/f$, two gain paths are matched, and the signal transfer function is straightened out to be linear as marked by the dotted line. However, if $1/f' > 1/f$, it is overcancelled, and the nonlinear error of the opposite polarity results. On the other hand, if $1/f' < 1/f$, the error is undercancelled. If $1/f' = 0$, no cancellation occurs at all. That is, the cancellation

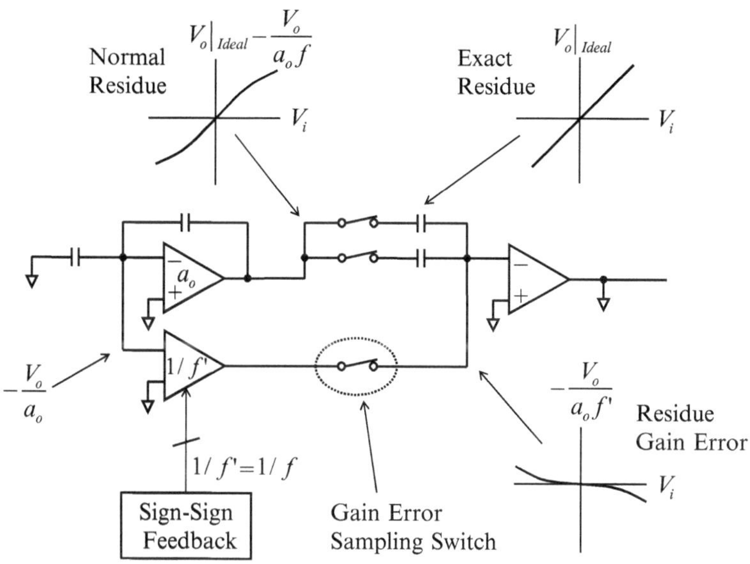

Fig. 5.9 Opamp gain and nonlinearity cancellation concept

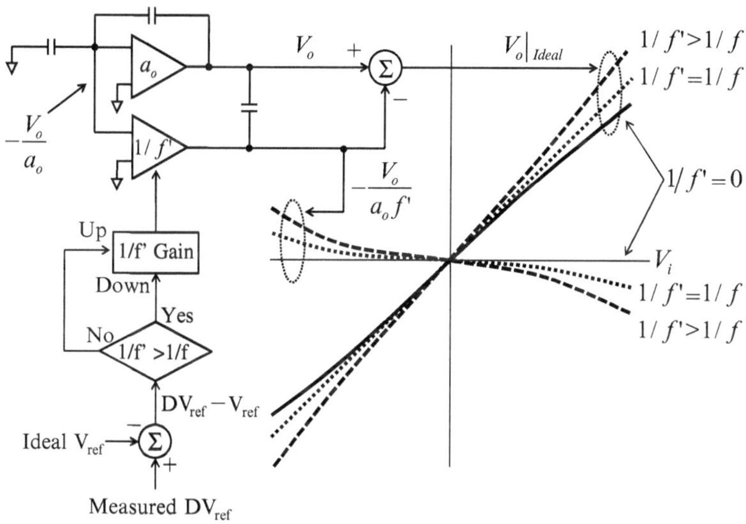

Fig. 5.10 Opamp gain and nonlinearity cancelling mechanism

depends on this single parameter, and the self-trimming process can be greatly simplified. As the estimated gain of $1/f'$ converges to the actual gain of $1/f$ over time, the transfer function approaches the dotted ideal straight line asymptotically.

5.6.1 Summing-Node Error Amplifier with Programmable Gain

The main opamp can be of any type of opamps, but the summing-node error amplifier should be a low-gain wideband amplifier with gain programmability to track the main amplifier summing-node gain. The default candidate is a differential pair as shown along with a main opamp in Fig. 5.11.

The MDAC opamp is made of a plain Miller-compensated two-stage opamp with no cascoding and gain-boosting. The uncascoded output stage works at low supply and attains the maximum output swing. The gain of the MDAC opamp is about 50–60 dB depending on the process and bias condition. Typically the achievable gain from two-stage differential pairs is not high enough to generate a 12b-level accurate residue with a closed-loop gain of 4–8, which is a typical requirement for residue amplifiers in high-resolution pipelined ADCs.

The error amplifier is a diode-loaded standard differential pair monitoring the opamp summing node as shown in Fig. 5.12. Its gain is digitally controlled by adjusting the bias current. From simple error analysis, the incremental gain step of the error amplifier can be derived as follows.

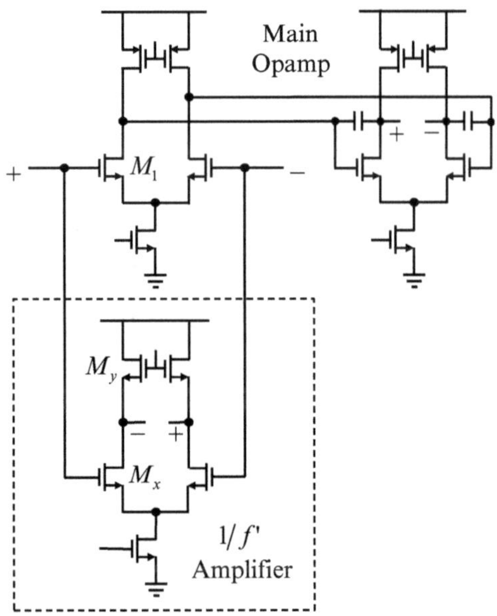

Fig. 5.11 Diode-loaded differential pair for the summing-node amplifier

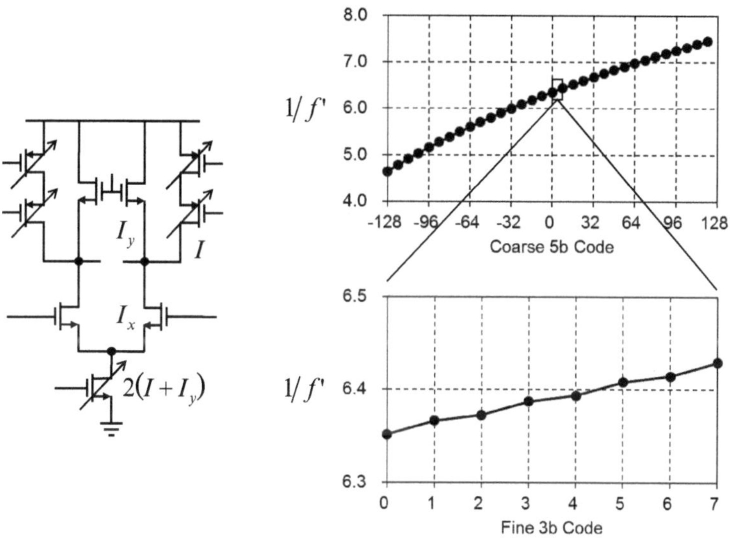

Fig. 5.12 Gain programming for the summing-node error amplifier

$$\Delta\left(\frac{1}{f'}\right) \approx \frac{1}{f'} \times \left(\frac{I}{I+I_y}\right) \times \left(\frac{\Delta W}{2W}\right), \tag{5.15}$$

where I and W are the bias current and the device width of the programmable current source, respectively. The bias current of M_x is programmed while that of the load M_y is fixed. If the nominal bias current of M_x is set to be about nine times higher than I_y, $I = 8I_y$, and programming I with 8b control using (5.15) results in a monotonic very fine gain step of about 0.01.

Considering the loading DAC capacitance C_{DAC2} of the second-stage MDAC, the frequency response of the error amplifier can be approximated with a dominant output pole as

$$\frac{1}{f'}(s) \approx \frac{g_{mx}}{\left(g_{my} + \frac{1}{r_{ox}} + \frac{1}{r_{oy}}\right)} \times \frac{1}{\left(1 + \frac{sC_{DAC2}}{g_{my}}\right)} \approx \frac{g_{mx}}{g_{my}} \times \frac{1}{\left(1 + \frac{s}{\omega_p}\right)}, \tag{5.16}$$

where g_{mx} and g_{my} are the transconductances, and r_{ox} and r_{oy} are the output resistances of transistors M_x and M_y, respectively. The programmable gain of $1/f'$ can be implemented by adjusting the bias currents that vary the g_m ratio.

A standard pipelined ADC with its first-stage opamp gain and nonlinearity adaptively self-trimmed using a programmable summing-node error amplifier is shown in Fig. 5.13.

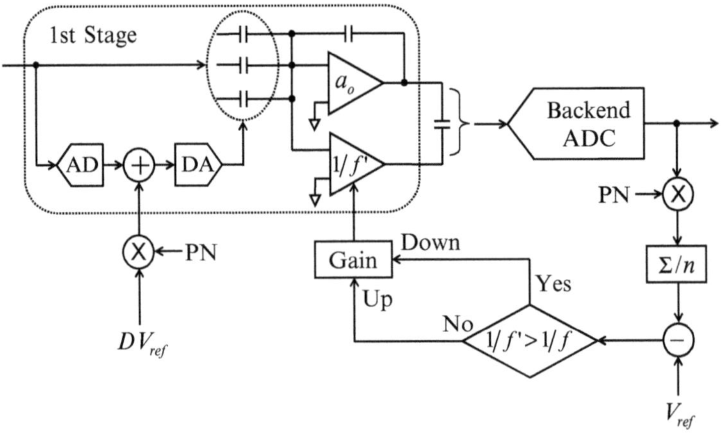

Fig. 5.13 Pipelined ADC with its first-stage calibrated

5.6.2 Gain Mismatch Polarity Detection by Digital Dithering

The sign–sign servo-feedback based on the LMS algorithm forces the gain mismatch to converge to zero by trimming the gain of the summing-node error amplifier. Digital dithering is to inject random dithers by shifting the coarse comparator threshold voltages up and down like a pseudo-random (PN) sequence of digital +1 and −1. Then the PN-modulated voltage can be recovered using the same PN dither sequence. The polarity of the gain mismatch error can be detected by comparing the recovered dither with the ideal dither digitally.

The first step is to correlate the digital output from the back-end ADC with the same PN sequence and to average the PN-demodulated output over many samples. Then this averaging is equivalent to measuring the dithered reference step $\mathrm{DV}_{\mathrm{ref}}$ at a very high oversampling rate. The error polarity is detected by comparing $\mathrm{DV}_{\mathrm{ref}}$ with its ideal value V_{ref}, and the up/down counter register that keeps the current digital gain setting of $1/f'$ is incrementally updated after many samples are averaged. Therefore, the accuracy of this LMS adaptation is only limited by the step size of the gain adjustment.

The residue output of the digitally dithered N-b MDAC can be written as

$$V_{\mathrm{res}} = 2^N \times V_{\mathrm{i}} - \left(\sum_{i=1}^{N} 2^{i-1} b_i + \frac{\mathrm{PN}}{2} \right) \times V_{\mathrm{ref}}, \tag{5.17}$$

where PN is +1 or −1, and b_i is the sub-ADC output. So the dithered output alternates between $+V_{\mathrm{ref}}/2$ and $-V_{\mathrm{ref}}/2$, and gives the actual digitized $\mathrm{DV}_{\mathrm{ref}}$, which is compared with the ideal V_{ref} for polarity detection. In general, the digital dithering is ineffective for nonlinearity measurements since noticeable high-order

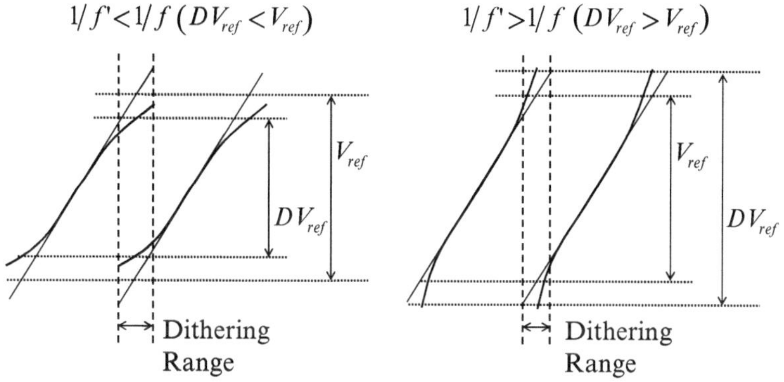

Fig. 5.14 Residue plots at a PN-dithered comparator threshold

distortions arise only with the actual input of large magnitude applied. However, the mismatch between two linear path gains is independent of the input magnitude, and the digital dithering suffices to measure one parameter of the gain mismatch. In the residue plot, the residue makes an abrupt shift-down/-up transition equivalent to V_{ref} at comparator thresholds, which is an analog processing to subtract/add a reference voltage to keep the residue inside the input range of the next stage.

Therefore, by changing the comparator threshold randomly, digital dithers can be injected as sketched in Fig. 5.14. Both the undercancelled and overcancelled cases are shown on the left and right sides, respectively. There exist two residue outputs within the dithering range, which covers the input between two adjacent PN-dithered comparator threshold voltages. The vertical gap $\mathrm{DV_{ref}}$ in the residue plot within the dithering range is matched to the ideal V_{ref} if the error polarity can be detected based on the following relation.

$$\mathrm{Sign}\left(\frac{1}{f'}-\frac{1}{f}\right)=\mathrm{Sign}(\mathrm{DV_{ref}}-V_{\mathrm{ref}}). \tag{5.18}$$

Note that when dithered digitally, the nonlinearity measurement is most effective when the input is close to the comparator thresholds, where the gain and nonlinearity error peaks and the comparator threshold can be dithered most effectively.

5.6.3 Self-Trimming Sequence

The error measurement cycle is to correlate out the PN-dithered error while de-correlating the signal. Examples of three possible digitally dithered residue plots are shown in Fig. 5.15.

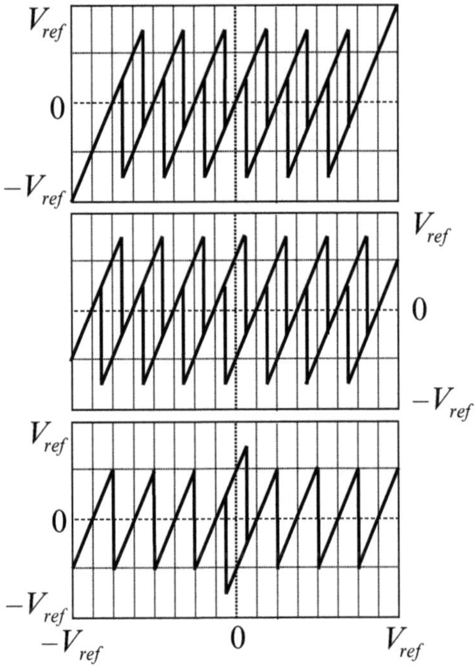

Fig. 5.15 Three examples of the 3b residue plot with digital dithering

They are the transfer function of the MDAC with comparator thresholds included. The top one is for the half-bit shifted 3b residue using the tri-level DAC, and the middle one is for the normal 3b differential MDAC. All comparator thresholds are dithered in the first two residue plots while in the last one, which is the same as the middle one, only one comparator threshold placed at the center is dithered. Also note that the last single-level dithering at zero-crossing offers the fastest convergence since only the nonlinearity error can be accumulated without de-correlating the large signal.

Assume that the dithering range is $\pm V_{\text{ref}}/16$ in the last residue plot of Fig. 5.15. When power is turned on, a fixed but unknown voltage within the dithering range is applied to the input, and n consecutive digital outputs are accumulated after correlated with the same PN sequence. Then from (5.17), the dither with 16b nonlinearity can be correlated out as follows.

$$\begin{aligned}\sum(\text{PN} \times V_{\text{res}}) &= \sum\left[\text{PN} \times \left\{2^2 \times V_{\text{i}} - \text{PN} \times \frac{V_{\text{ref}}}{2}\left(1 \pm \frac{1}{2^{16-2}}\right)\right\}\right] \\ &\approx -n \times \frac{V_{\text{ref}}}{2}\left(1 \pm \frac{1}{2^{14}}\right),\end{aligned} \tag{5.19}$$

where V_{i} is amplified by the MDAC gain of 2^2. Uncorrelated terms are randomized and averaged out to be zero. For the measured dither in (5.19) to be larger than the residual input, the minimum number of samples to complete one error measurement is given by

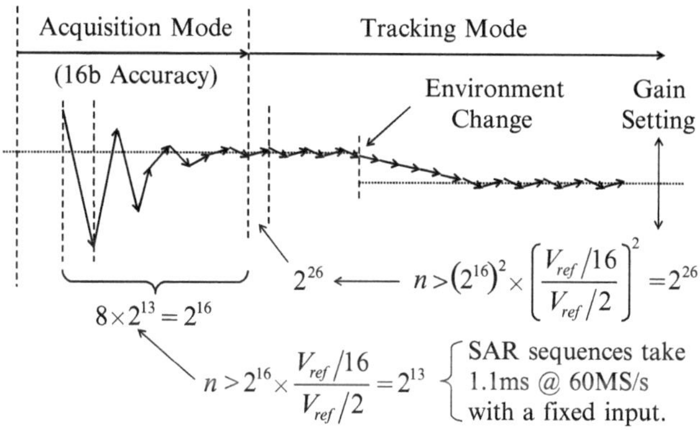

Fig. 5.16 Initial adaptation cycles

Fig. 5.17 Signal de-correlation vs. number of sampling

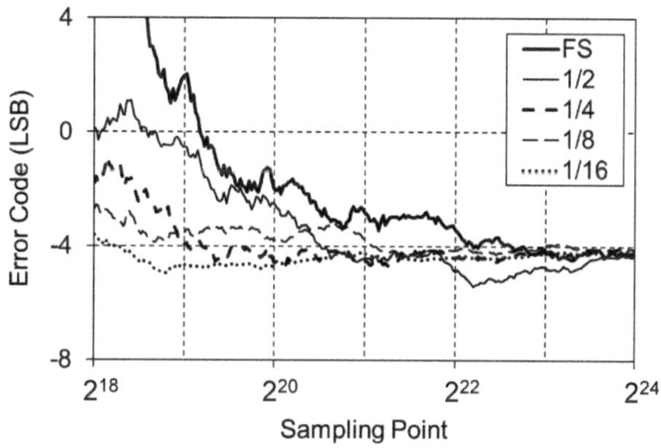

$$n > 2^{14} \times \frac{2^2 \times V_{\text{ref}}/16}{V_{\text{ref}}/2} = 2^{13}, \tag{5.20}$$

where $V_{\text{ref}}/16$ is the worst-case de-correlated average of the fixed input V_{i}.

As explained in Fig. 5.16, if an 8b current DAC is used to trim this gain, about eight times more samples of 2^{16} are enough to complete the eight cycles of simple successive approximation. It only takes 1.1 ms for the initial 16b-accurate acquisition if sampled at 60 MS/s. In practice, it requires far fewer samples than 2^{16} since the fixed DC input can be averaged out to be almost zero and leave no residual term.

Figure 5.17 shows how the random signal is de-correlated. In the tracking mode, the measurement cycle is significantly lengthened since the random AC input should be de-correlated and the correlation accuracy improves only as a square root function of the number of samples. If the same input average of $V_{\text{ref}}/16$ is applied while all comparator levels are dithered, the minimum number of samples to make one decision is $(2^{13})^2 = 2^{26}$, and it takes 1.1 s at 60 MS/s. In reality, since the error correlation works only when the input falls into the dithering range, the polarity decision can be delayed further.

It takes far more samples to de-correlate the largest full-scale signal. As is true in digital calibration schemes, it takes prohibitively long time to calibrate the opamp gain and nonlinearity error with the random AC input applied. To shorten the measurement cycle, it is desirable to keep the signal-to-dither ratio as small as possible. However, the best solution to shorten the initial acquisition time is the power-on calibration performed either with the input briefly shorted or kept constant. After the initial acquisition, any slow parameter drift can be tracked with the random input applied. For this PN-dithering, the standard extra half-bit redundancy range for digital correction is reduced to make room for the dither. Therefore, comparators need to resolve one extra bit to contain the residue output inside the digital correction range.

5.6.4 Accuracy Considerations

If a high-gain opamp is used as an MDAC amplifier, its limited output range usually distorts the DC transfer curve. Since it can be cancelled collectively as discussed, it is not necessary to elaborate further on the MDAC nonlinearity in detail, and a simple low-gain open-loop amplifier can be used as a summing-node error amplifier. A single-stage resistor-loaded differential pair suffers from the nonlinearity of the input device. If an active load is used, the nonlinearity of the load also contributes to the total error.

If any voltage V is distorted, it can be modified in a normalized form of $V+\alpha V^3$ including the dominant cubic error term. This third-order nonlinearity term of αV^3 can be estimated either at the input or at the output if the coefficient α is scaled accordingly. That is, the nonlinearity of the auxiliary $1/f'$ amplifier can be modeled as shown in Fig. 5.18, where $\alpha_{1/f'}$ is the third-order nonlinearity coefficient.

After the condition of $1/f' = 1/f$ is reached by feedback, the residue output becomes

$$V_{\text{res}} = V_{\text{o}}\big|_{\text{Ideal}} + \alpha_{1/f'} \times \left(\frac{V_{\text{o}}}{a_{\text{o}} f'}\right)^3. \tag{5.21}$$

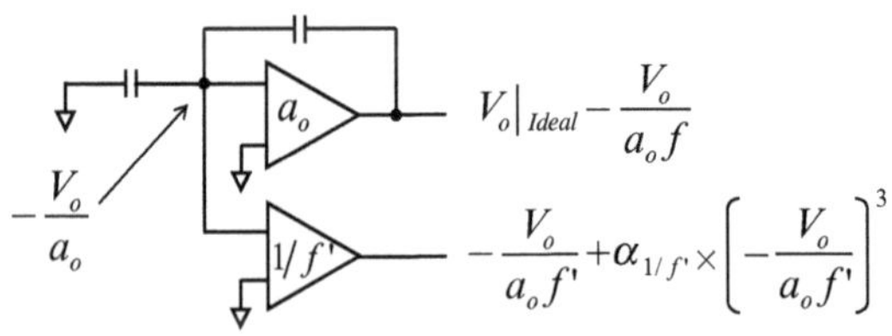

Fig. 5.18 Summing-node error amplifier with third-order nonlinearity

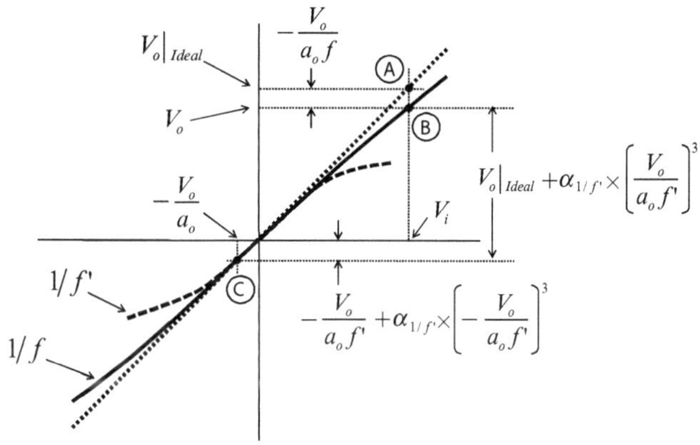

Fig. 5.19 Nonlinearity of the summing-node amplifier in the transfer curve

The residue transfer accuracy of the switched-capacitor MDAC can be also analyzed using its DC transfer curve. Three superposed transfer curves from the opamp summing node to the outputs are shown in Fig. 5.19.

The dotted line is the ideal one, and the solid and dashed lines are those of the MDAC and the auxiliary amplifier, respectively. Ideally, the MDAC output should settle to the point A for the given input V_i and yield the ideal residue output $V_o|_{Ideal}$ set only by the capacitor ratio. In practice, it settles to the point B to yield the output V_o, which is slightly lower than the ideal $V_o|_{Ideal}$. The difference is $-V_o/a_of$ which represents the total opamp-induced error. This error also appears as $-V_o/a_o$ at the summing node after attenuated by the feedback factor f. If the error amplifier is linear, the amplified summing-node error becomes $-V_o/a_of'$, which cancels $-V_o/a_of$ at the point C. The ideal residue output of $V_o|_{Ideal}$ is the vertical distance between two points B and C.

The worst-case residue error occurs when the output is the full-range V_{ref}. This nonlinearity error in the residue output is negligible if it is smaller than one LSB of the back-end ADC by meeting the following linearity requirement.

$$\alpha_{1/f'}\left(\frac{V_{ref}}{a_o f'}\right)^3 < \frac{V_{ref}}{2^{14}}, \tag{5.22}$$

if the back-end ADC of this example resolves 14b. In high-resolution ADCs, the output range of the $1/f'$ amplifier is much smaller than that of the MDAC since it is reduced by the large DC loop gain of a_of. As a result, the linearity requirement of the $1/f'$ summing-node error amplifier is far less stringent than that of the MDAC.

Consider the following example. $V_o|_{Ideal} = 1$ V, $a_o = 1000$, and $f = 1/5$. We obtain the loop gain of $a_of = 200$ and the gain error of $V_o/a_of = 5$ mV, and the actual MDAC residue output becomes $V_{res} = 1\ \text{V} - 5\ \text{mV} = 0.995$ V. Now if the nonlinearity error is assumed to be 50 μV, the amplified summing-node error decreases slightly to $-5\ \text{mV} + 50\ \mu\text{V}$, and the residue to be sampled by the next stage becomes $V_{res} = 1\ \text{V} - 50\ \mu\text{V} = 0.99995\ \text{V}$. Thus it implies that the 50 μV nonlinearity error is about −40 dB of the 5 mV summing-node amplifier output, but when referred to the 1 V residue output, it is a small error of −86 dB improved by the loop gain of 200. The linearity requirement of the summing-node error amplifier is greatly alleviated by increasing the loop gain. However, if simple open-loop or low loop-gain residue amplifiers are used as an MDAC, the linearity requirement for the error amplifier gets a little stringent due to the reduced amount of the loop gain.

In ADCs, any residual error after cancelled can be detected digitally. Open-ended digital calibration methods sort out errors after they occur, and the effectiveness of calibration heavily relies on the error measurement accuracy. On the other hand, close-ended adaptive feedback approaches can eliminate the error at its source in the analog domain. It is the standard zero-forcing LMS feedback that performs such tasks perfectly. It requires no additional post analog or digital processing since the opamp-induced error is eliminated before sampled. Using only the polarity of the detected residual error, the zero-forcing algorithm forces the amplifier gain error to converge to zero. This is in effect close to the physical trimming of characteristics of electronic components. If done electronically, self-trimming rather than self-calibration is the proper word for this process. Since the DC servo feedback is based on oversampling, the achievable accuracy is not limited by the measurement accuracy but by the step size for the gain trim. Therefore, there is no need for extra bits of resolution in the error measurement that other digital calibration methods require.

5.7 Noise Implication of Nonlinearity Cancellation

Since the noise at the opamp summing node is amplified in two different paths with the same gain and subtracted later, the proposed nonlinearity cancellation has the same noise implication as in the noise-cancelling LNAs as shown in Fig. 5.20.

Although the cancellation works for all the in-band low-frequency noise and offset of the main opamp as they are referred to the same summing node, those of the summing-node error amplifier remain. The main thermal noise sources of two gain paths and their band-limited output spectral densities are sketched in Fig. 5.21.

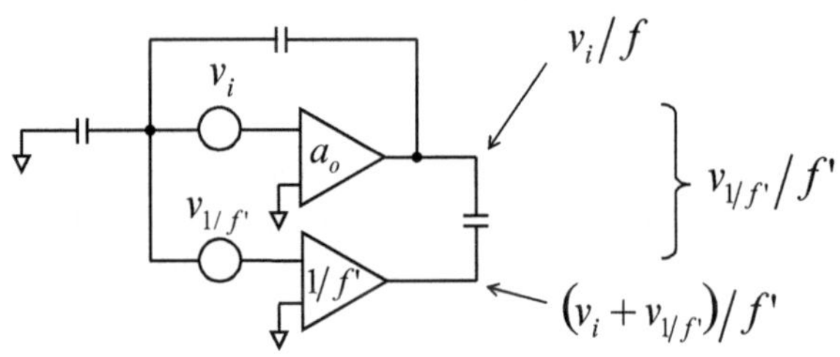

Fig. 5.20 Noise-cancelling mechanism

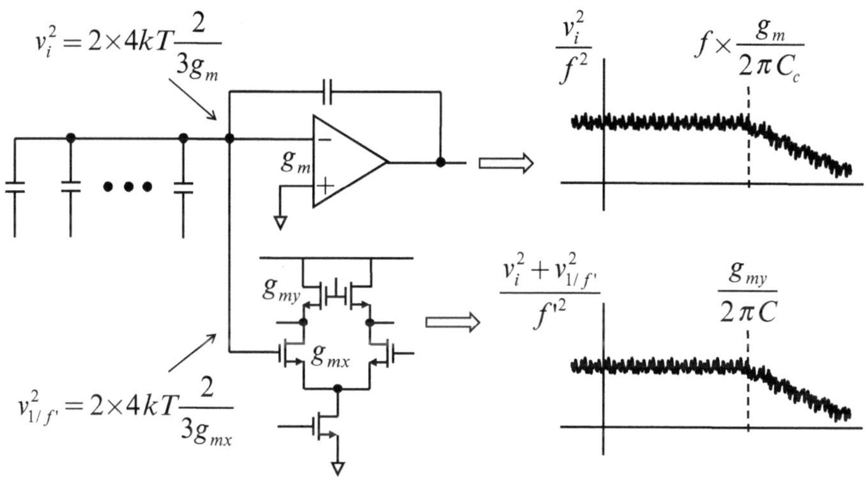

Fig. 5.21 Band-limited noises of two gain paths

The value of the sampling capacitor is the most traded parameter for speed and power in switched-capacitor amplifier designs. If the total input capacitance of the first-stage MDAC is C_{DAC1}, the differential bottom-plate *S*/*H* samples the *kT*/*C* noise of

$$v_{kT/C}^2 = 2 \times \frac{kT}{C_{\mathrm{DAC1}}}. \tag{5.23}$$

Once sampled, this noise is added to the signal, and usually sets the lower bound of the dynamic range. The input-referred thermal noise of the main opamp is also band-limited. If the input differential pair is assumed to be the dominant noise source, the input referred in-band thermal noise power is estimated as

$$v_{\mathrm{i}}^2 = 2 \times 4kT\frac{2}{3g_{\mathrm{m}}} \times f\frac{g_{\mathrm{m1}}}{2\pi C_{\mathrm{c}}} \times \frac{\pi}{2} = 2 \times \frac{kT}{C_{\mathrm{c}}} \times \frac{2}{3}f, \tag{5.24}$$

where C_{c} is the Miller compensation capacitance of the two-stage opamp, $f(g_{\mathrm{m}}/C_{\mathrm{c}})$ is the closed-loop bandwidth, and $\pi/2$ is the common factor to take the out-of-band noise power above the one-pole roll-off frequency into account. Similarly, the input-referred in-band thermal noise of the error amplifier is

$$v_{1/f'}^2 = 2 \times 4kT\frac{2}{3g_{\mathrm{mx}}} \times \frac{g_{\mathrm{my}}}{2\pi C_{\mathrm{DAC2}}} \times \frac{\pi}{2} = 2 \times \frac{kT}{C_{\mathrm{DAC2}}} \times \frac{2}{3}f', \tag{5.25}$$

where g_{my} is the diode load resistance and C_{DAC2} is the input capacitance of the second-stage MDAC. The band-limited in-band noises of both the residue amplifier

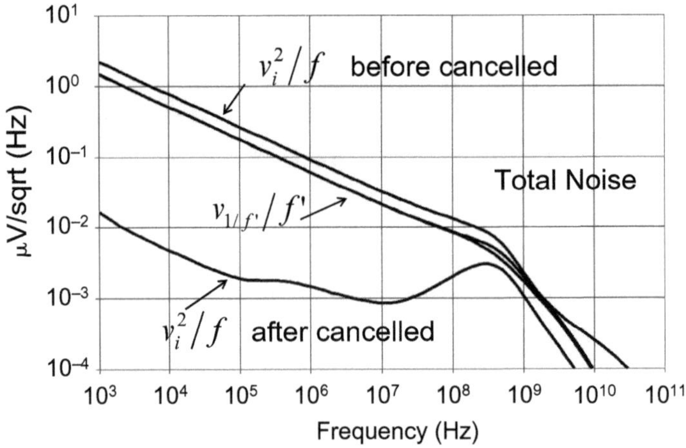

Fig. 5.22 Simulated noises of two gain paths

and the error amplifier are similar to the kT/C noise when sampled by the next stage, but they are scaled down by $2f/3$ and $2f'/3$, respectively.

Noise spectral densities are simulated to give the total noise of the residue output to be sampled by the next stage in Fig. 5.22. The total output noise approaches that of the error amplifier within the bandwidth as the opamp noise of the MDAC is cancelled [12]. Therefore, the total noise can be designed to be lower if the condition of $C_{\mathrm{DAC2}} > C_{\mathrm{c}}$ is met. The input devices of the error amplifier can be made large with long channels for low flicker noise. That is, it is easier to achieve low power and low noise even with both a low-gain MDAC opamp and an extra error amplifier than with just one high-gain MDAC opamp. The MDAC opamp can be designed to consume less power because its noise is cancelled and no gain-boosting for high gain is required.

5.8 Effect of High-Frequency Zero on Settling

In all parallel or feedforward systems, a feedforward zero is created in their transfer function. If the MDAC opamp has a bandwidth of $\omega_{-3\mathrm{dB}}$, the closed-loop MDAC output is modified as follows.

$$V_{\mathrm{o}}(s) = \frac{V_{\mathrm{o}}(s)|_{\mathrm{Ideal}}}{1 + \frac{1}{a_{\mathrm{o}} f} \times \left(1 + \frac{s}{\omega_{-3\,\mathrm{dB}}}\right)} = \frac{V_{\mathrm{o}}(s)|_{\mathrm{Ideal}}}{\left(1 + \frac{1}{a_{\mathrm{o}} f}\right) \times \left(1 + \frac{s}{\omega_k}\right)}, \tag{5.26}$$

where the closed-loop unity-gain bandwidth ω_k is defined as

$$\omega_k = (1 + a_o f)\omega_{-3\,\text{dB}} \approx a_o f \times \omega_{-3\,\text{dB}}. \tag{5.27}$$

If the error amplifier has a bandwidth of ω_p from (5.16), the nonlinearity-cancelled residue output is now approximated as

$$\begin{aligned} V_\text{o}(s) - \frac{V_\text{s}}{a_o f'}(s) &= \left\{ 1 + \frac{1}{a_o f'} \times \frac{\left(1 + \dfrac{s}{\omega_{-3\,\text{dB}}}\right)}{\left(1 + \dfrac{s}{\omega_\text{p}}\right)} \right\} \times \frac{V_\text{o}(s)|_\text{Ideal}}{\left(1 + \dfrac{1}{a_o f}\right)\left(1 + \dfrac{s}{\omega_k}\right)} \\ &= \frac{\left(1 + \dfrac{1}{a_o f'}\right)\left(1 + \dfrac{s}{\omega_z}\right)}{\left(1 + \dfrac{1}{a_o f}\right)\left(1 + \dfrac{s}{\omega_k}\right)\left(1 + \dfrac{s}{\omega_\text{p}}\right)} \times V_\text{o}(s)|_\text{Ideal} \\ &\approx \frac{\left(1 + \dfrac{s}{\omega_z}\right)}{\left(1 + \dfrac{s}{\omega_k}\right)\left(1 + \dfrac{s}{\omega_\text{p}}\right)} \times V_\text{o}(s)|_\text{Ideal}, \end{aligned} \tag{5.28}$$

if $1/f' = 1/f$. The frequency response has two poles and one zero. Note that a zero is created at ω_z, which is approximately

$$\omega_z = \left(1 + \frac{1}{a_o f'}\right) \times \left(a_o f' \times \omega_{-3\,\text{dB}} || \omega_\text{p}\right) \approx \frac{1}{\frac{1}{\omega_k} + \frac{1}{\omega_\text{p}}} = \omega_k || \omega_\text{p}. \tag{5.29}$$

The zero doesn't affect the stability of the feedback system, but the transient response to the step input is greatly affected by the zero in the transfer function. Feedback amplifiers with low-Q poles only respond to the step input exponentially with phase-lagging while the zero adds the phase-leading differentiated step response. Three cases of transient responses are compared in Fig. 5.23.

Poles of most feedback opamps are complex conjugates. Their transient responses to the step input vary widely depending on the location of the complex conjugate poles. Low-Q poles are the most desirable since their step responses are exponential with no significant overshoot and ringing associated with high-Q poles. A zero improves the phase margin (PM) by drawing the conjugate poles away from the imaginary axis, but the zero introduces the phase-leading advance of the transient input. Therefore, compared to all-pole response, the step response peaks first and settles to the final value.

Due to the causality condition, the phase-leading differentiated response with an optimally placed zero helps settling when it partially cancels out the phase-lagging

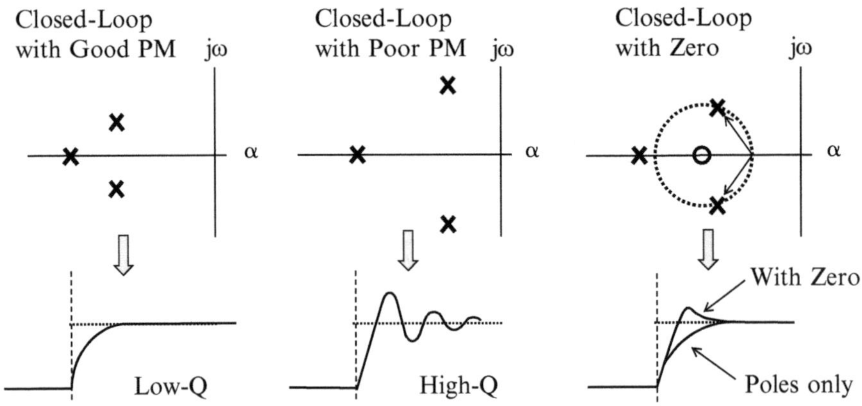

Fig. 5.23 Three complex poles and transient responses

component. This condition can be met only at specific conditions and output magnitudes, but overall it is true that the phase-leading zero shortens the settling process. The normalized transient response to the step input can be obtained by taking the inverse Laplace transform of (5.28).

$$\begin{aligned} v_o(t) - \frac{v_o}{a_o f'}(t) &= \frac{\omega_p}{\omega_p - \omega_k} \times \left(1 - e^{-\omega_p t}\right) - \frac{\omega_k}{\omega_p - \omega_k} \times \left(1 - e^{-\omega_k t}\right) \\ &= 1 + \frac{\omega_k}{\omega_p - \omega_k} e^{-\omega_k t} - \frac{\omega_p}{\omega_p - \omega_k} e^{-\omega_p t}. \end{aligned} \tag{5.30}$$

Note that if two poles of the MDAC and the error amplifier are matched ($\omega_p = \omega_k$), the output settles instantly. However, the inevitable mismatch between them gives rise to the classical doublet problem in settling. In practice, two poles are separated to avoid the slow settling by doublet. There exist three settling components. The first one is the ideal output, and the second one is the phase-leading term, and the third one is the phase-lagging term. Usually, switched-capacitor circuits with one dominant pole settle with the phase-lagging term of $\exp(-\omega_k t)$ only.

As shown in Fig. 5.24, the settling trajectory of the output is bounded in between two exponentially settling curves marked with two dashed lines, which are given by the leading and lagging components. As shown in (5.30), the phase-lagging term can settle fast with $\exp(-\omega_p t)$. After the phase-lagging term settles quickly, the phase-leading term settles if $\omega_p > \omega_k$. If $\omega_p \gg \omega_k$, the output settles fast since the zero at ω_k cancels ω_p. The dotted line represents the single-pole exponential settling without a phase-leading zero. The output passes the ideal residue output at t_z, and peaks at t_p as given by

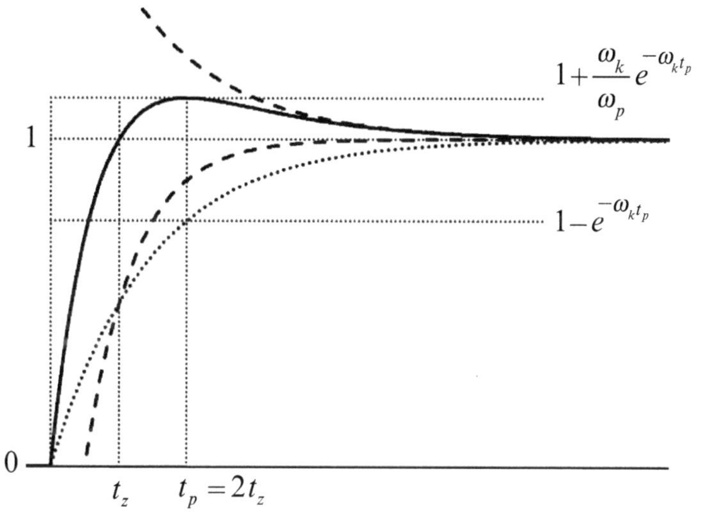

Fig. 5.24 Settling with two poles and a zero

$$t_z = \frac{\ln\left(\frac{\omega_p}{\omega_k}\right)}{\omega_p - \omega_k}, \quad \text{and} \quad t_p = 2t_z = \frac{2\ln\left(\frac{\omega_p}{\omega_k}\right)}{\omega_p - \omega_k}, \tag{5.31}$$

respectively. Therefore, its peak value is obtained from (5.30) and (5.31) as

$$\begin{aligned} V(t_p) &= 1 + \frac{\omega_k}{\omega_p}e^{-\omega_k t_p} - \frac{\omega_k}{\omega_p}e^{-\omega_k t_p} + \frac{\omega_k}{\omega_p - \omega_k}e^{-\omega_k t_p} - \frac{\omega_p}{\omega_p - \omega_k}e^{-\omega_p t_p} \\ &= 1 + \frac{\omega_k}{\omega_p}e^{-\omega_k t_p} + \frac{1}{\omega_p}\left(\frac{\omega_k^2}{\omega_p - \omega_k}e^{-\omega_k t_p} - \frac{\omega_p^2}{\omega_p - \omega_k}e^{-\omega_p t_p}\right) \\ &= 1 + \frac{\omega_k}{\omega_p}e^{-\omega_k t_p}. \end{aligned} \tag{5.32}$$

That is, due to the zero, the output approaches the final voltage faster than without the zero. After the output peaks, it settles back to the final voltage since this peaking gives an initial condition for MDAC settling assuming that the ω_p term has already settled.

However, in the nominal single-pole settling, the output $t = t_p$ is

$$V(t_p) = 1 - e^{-\omega_k t_p}. \tag{5.33}$$

This implies that the pole makes the output approach the final value closer at $t = t_p$ if $\omega_p > \omega_k$ is true. The single dominant pole settling time is defined as the time it takes for the error to exponentially decrease to below $1/a_o f$.

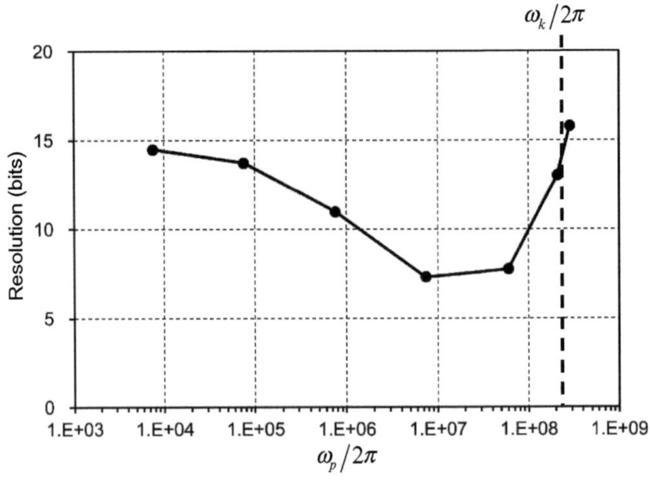

Fig. 5.25 Resolution of 3b MDAC vs. the error amplifier pole

$$t_s = \frac{1}{\omega_k}\ln(a_o f). \tag{5.34}$$

If $t = t_s$, the exponential decay term equals the constant error term due to the finite loop gain. If $\omega_p > \omega_k$, the following inequality condition warrants that (5.30) can settle faster than the single dominant pole case.

$$\frac{\omega_k}{\omega_p - \omega_k}e^{-\omega_k t_s} - \frac{\omega_p}{\omega_p - \omega_k}e^{-\omega_p t_s} < e^{-\omega_k t_s}. \tag{5.35}$$

For (5.35) to be true, the condition of $\omega_p = \omega_k$ or $\omega_p > 2\omega_k$ should be met.

A switched-capacitor 3b MDAC in the pipeline ADC is simulated as shown in Fig. 5.25. Note that the amplifier settles accurately to give high-resolution residues if either one of the conditions of $\omega_p \gg \omega_k$ and $\omega_p \ll \omega_k$ is met. However, the latter has no practical meaning.

The effect of the zero on the normalized step response is simulated with the MDAC bandwidth ω_k normalized to 250 MHz as shown in Fig. 5.26. The higher the ratio of ω_p/ω_k gets, the sharper the peak of the phase-leading component gets. In practice, the zero given in (5.29) is placed anywhere from $\omega_k/2$ to ω_k if the bandwidth of the error amplifier is set wider than that of the MDAC. The summing-node error amplifier can be easily designed for this since the swing of its output is much smaller than that of the opamp. The end result is that the MDAC can settle fast and more accurately with its noise cancelled while the linearity requirement is greatly reduced.

Figure 5.27 shows the simulated transient outputs of the MDAC, the error amplifier, and the summing node. The main amplifier output settles inaccurately due to the opamp gain error, but the difference between the MDAC output and the amplified summing-node error makes an ideal residue output. In simulations, if ω_p

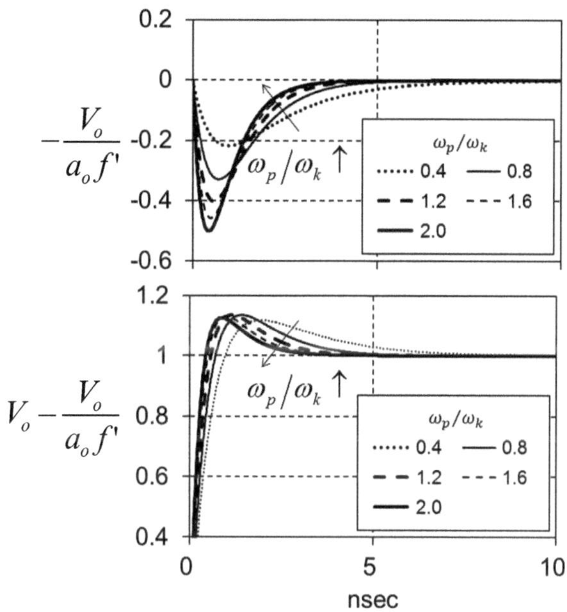

Fig. 5.26 Zero effect on settling

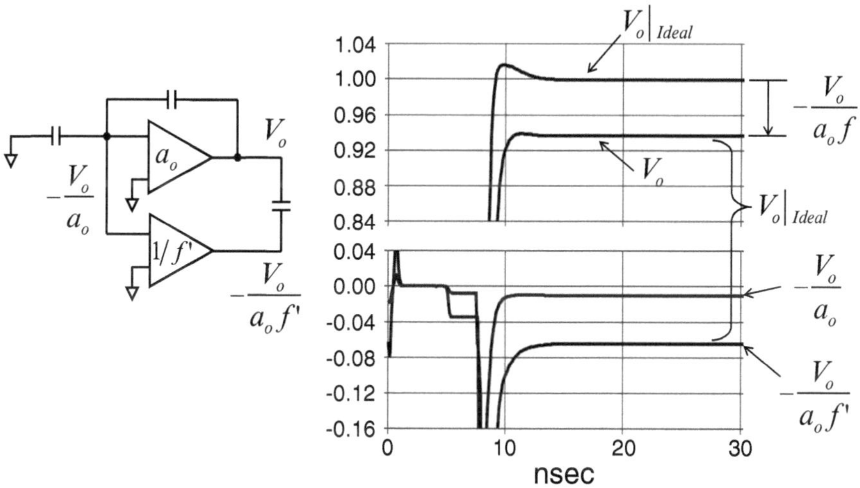

Fig. 5.27 Simulated normalized step response

is set to be 20 % higher than ω_k, the peaking effect is largely mitigated, and the residue amplifier settles with a dominant time constant of $1/\omega_k$ with no significant overshoot. If the input is large, the MDAC output will slew nonlinearly at the beginning due to the sharp input spike exceeding its linear range, but the high slew-rate will move the circuit back into the linear mode quickly in most high-resolution switched-capacitor circuits.

5.9 Experimental Results

A 14b pipelined ADC based on a four-capacitor tri-level MDAC resolves 3b per stage without using the input *S*/*H*. Both the first-stage MDAC and comparators sample the input simultaneously during the split interval of the sampling phase, and coarse ADCs make decisions during the rest of the sampling phase [13]. To reduce the switch-related sampling nonlinearity, all sampling clocks are boosted, and capacitors are matched to generate a 12b residue [14, 15]. The nonlinearity-cancelled bottom-plate sampling realizes accurate interstage residue transfer, and alleviates the stringent requirement in the design of high-gain wideband opamps. The total opamp-induced error is removed using an opamp input summing-node error monitoring algorithm, which also eliminates the opamp noise and offset. Two key circuit concepts are required. One is to make an analog programmable gain element that adjusts the opamp gain and nonlinearity, and the other is to implement a digital oversampling quantizer that detects the error polarity with high precision.

The core functional blocks can be integrated on 2.2 mm × 0.65 mm in 0.18 μm CMOS. The total input capacitance is 6 pF. The chip consumes 68 mW from 1.6 V at 60 MS/s, which are divided into 39 mW in the ADC, 18 mW in the clock, and 11 mW in the voltage reference and bias circuits. The first-stage MDAC and its summing-node error amplifier consume 20 and 7.4 mW, respectively. The error amplifier consumes about 10 % of the total power. The full-scale signal range is $2V_{pp}$ differential. All measured results can be obtained with no capacitor calibration.

Figure 5.28 explains how the zero-forcing servo-feedback algorithm actually finds the optimum gain setting by successive approximation. The measured SNR, SNDR, and SFDR peak when the two-path gains are matched. Note that the polarity of the gain error changes at the optimum setting. This error polarity change facilitates the successive approximation algorithm shown on the right side to find the correct gain setting quickly. The gain control code for $1/f'$ is updated after checking 4k samples with a fixed DC input, and it takes about 0.5 ms to collect 32k samples for initial acquisition. The whole adaptation loop converges as shown with over 16b accuracy, which is limited by the finite trimming step for the g_m control.

Figures 5.29 and 5.30 show the INL and FFT plots measured at 60 MS/s before and after cancellation. Both the DNL and INL are smaller than ±0.6 LSB at 14b. With 1 MHz input, the SNR, SNDR, and SFDR are measured to be 78.5, 76.9, and 91.2 dB, respectively. With 31 MHz input, they are slightly degraded to be 75.5, 73.3, and 84.0 dB, respectively. The measured static linearity approaches the 15b level. The dynamic performance at the Nyquist rate is shown in Fig. 5.31 by sampling 1 and 31 MHz inputs at varying sampling rates ranging from 20 to 100 MS/s.

The SFDR peaked at 95 dB, but the dynamic performance was degraded gradually as typically measured in pipelined ADCs. Residual nonlinearity errors result from the trim step size of the gain $1/f'$, the capacitor mismatch, and the back-end ADC resolution. The $1/f'$ amplifier gain converges to about 6.

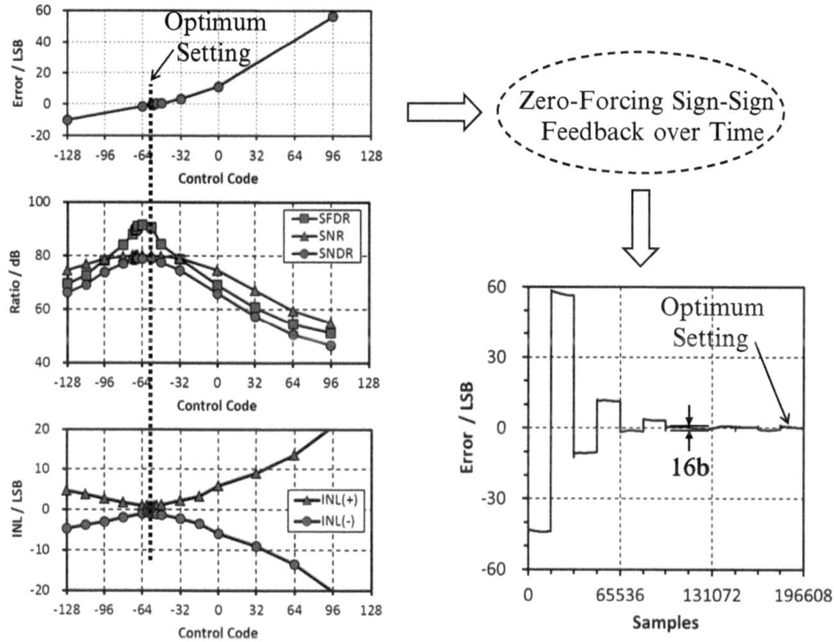

Fig. 5.28 Adaptive self-trimming based on zero-forcing LMS algorithm

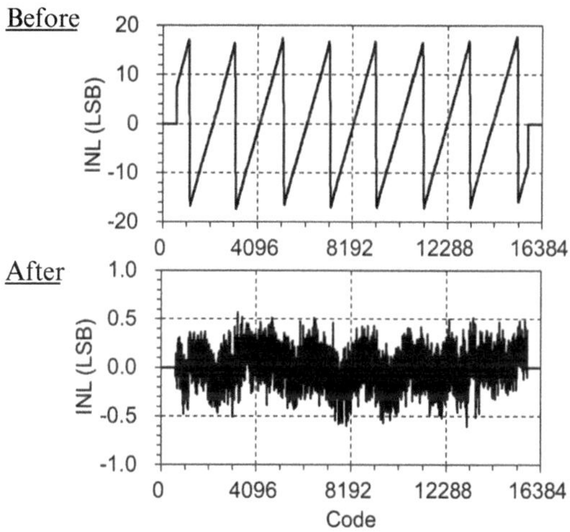

Fig. 5.29 INL at 14b before and after self-trimming

Figure 5.32 compares INLs measured from two ADCs at different supply voltages. One is the self-trimming version cancelling opamp gain and nonlinearity, and the other is a standard opamp version with a cascoded first stage consuming the same power as the self-trimming one. Note also that the DC gain of the conventional cascoded opamp is not high enough to give the required linearity at 14b.

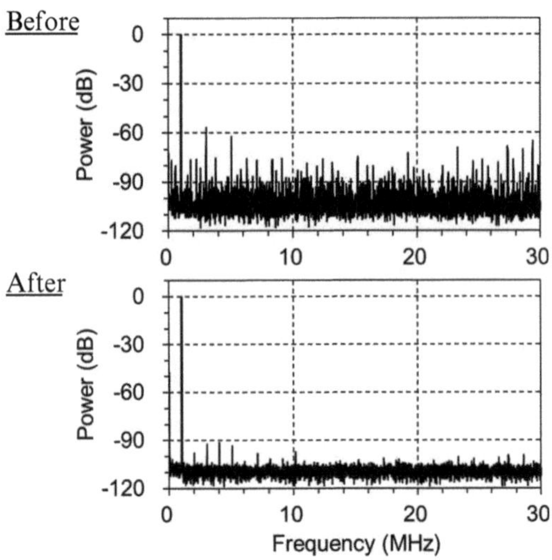

Fig. 5.30 FFT before and after self-trimming

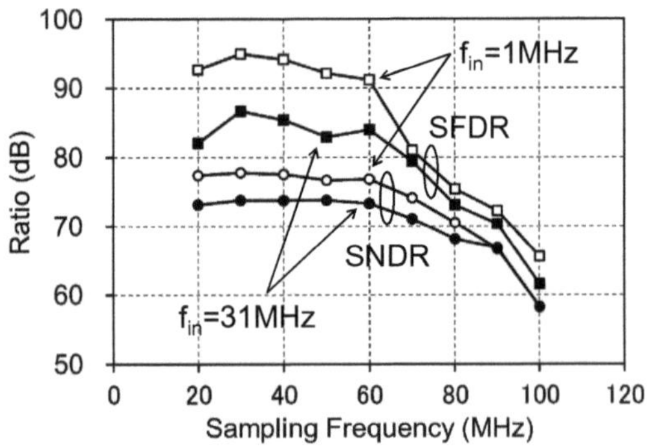

Fig. 5.31 SNDR and SFDR vs. f_s and f_{in}

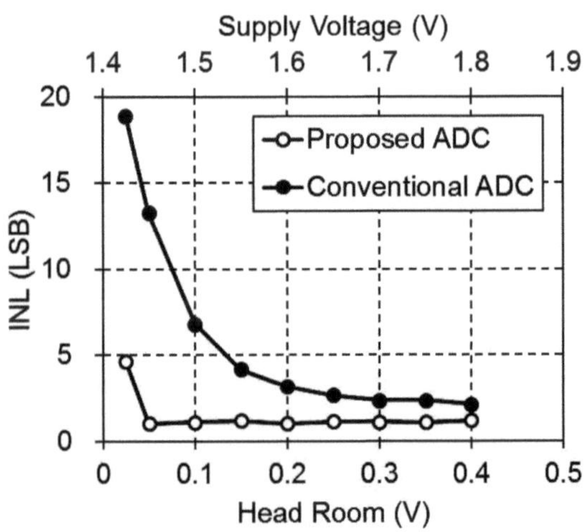

Fig. 5.32 INL vs. supply voltage and headroom

The self-trimming approach maintains high linearity down to 1.45 V supply with a headroom of only 50 mV per each transistor.

Experimental results validate the adaptive self-trimming concept that eliminates the gain and nonlinearity error of the MDAC-based residue amplifier. The accurate interstage voltage transfer is realized by adjusting the opamp-induced error based on sign–sign zero-forcing LMS feedback. In effect, adjusting a single gain parameter is equivalent to straightening the whole MDAC transfer function, and thereby the error-free residue can be generated. The adaptive approach can facilitate the design of high-SFDR pipelined ADCs to a great extent without requiring stringent opamp gain requirements.

References

1. B. Song, M. Tompsett, K. Lakshmikumar, A 12-Bit 1-Msample/s capacitor error averaging pipelined A/D converter. IEEE J. Solid State Circuits **SC-23**, 1324–1333 (1988)
2. B. Song, S. Lee, M. Tompsett, A 10b 15MHz CMOS recycling two-step A/D converter. IEEE J. Solid State Circuits **SC-25**, 1328–1338 (1990)
3. A. Ali, K. Nagaraj, Background calibration of operational amplifier gain error in pipelined A/D converters, IEEE Trans. Circuits Syst. II **50**, 631–634 (2003)
4. S. Sin, S. U, R. Martins, A novel low-voltage finite-gain compensation technique for high-speed reset- and switched-opamp circuits, in *Proceedings of IEEE International Symposium on Circuits and Systems*, May 2006, pp. 3794–3797
5. A. Ali et al., A 16b 250MS/s IF-sampling pipelined ADC with background calibration. IEEE J. Solid State Circuits **SC-45**, 2602–2612 (2010)
6. B. Murmann, B. Boser, A 12 b 75 MS/s pipelined ADC using open-loop residue amplification. IEEE J. Solid State Circuits **SC-38**, 2040–2050 (2003)
7. J. Keane, P. Hurst, S. Lewis, Background interstage gain calibration technique for pipelined ADCs. IEEE Trans. Circuits Syst. I **52**, 32–43 (2005)
8. A. Panigada, I. Galton, A 130 mW 100 MS/s pipelined ADC with 69dB SNDR enabled by digital harmonic distortion correction. IEEE J. Solid State Circuits **SC-44**, 3314–3328 (2009)
9. Y. Miyahara, M. Sano, K. Koyama, T. Suzuki, K. Hamashita, B. Song, Adaptive cancellation of gain and nonlinearity errors in pipelined ADCs, in *Digests of Technical Papers, IEEE International Solid-State Circuits Conference*, Feb. 2013, pp. 282–283
10. P. Li, M. Chin, P. Gray, R. Castello, A ratio-independent algorithmic analog-to-digital conversion technique. IEEE J. Solid State Circuits **SC-19**, 828–836 (1984)
11. Y. Miyahara, M. Sano, K. Koyama, T. Suzuki, K. Hamashita, B. Song, A 14b 60MS/s pipelined ADC adaptively cancelling opamp gain and nonlinearity. IEEE J. Solid State Circuits **50**, 416–425 (2014)
12. F. Bruccoleri, E. Klumperink, B. Nauta, Noise cancelling in wideband CMOS LNAs, in *Digest of Technical Papers, IEEE International Solid-State Circuits Conference*, Feb. 2002, pp. 406–407
13. I. Mehr, L. Singer, A 55 mW, 10 bit, 40 Msample/s Nyquist-rate CMOS ADC. IEEE J. Solid State Circuits **SC-35**(3), 318–325 (2000)
14. T. Brooks, D. Robertson, D. Kelly, A. Del Muro, S. Harston, A Cascaded sigma-delta pipeline A/D converter with 1.25-MHz signal bandwidth and 89dB SNR. IEEE J. Solid State Circuits **SC-32**, 1896–1906 (1997)
15. A. Abo, P. Gray, A 1.5-V, 10-bit, 14.3-MS/s CMOS pipelined analog-to-digital converter. IEEE J. Solid State Circuits **SC-34**, 599–606 (1999)

Chapter 6
RF Circuits

There are two types of high-frequency circuits. One is a wideband circuit covering DC to RF or microwave frequencies. The other is a narrowband circuit operating at RF or microwave frequencies. The former is broadbanded by feedback while the latter operates in open loop, but occasionally with local feedback. The former is also for wireline-baseband systems such as fiber and networking, but the latter is mostly for wireless RF transceivers. The key RF circuit elements are low-noise amplifier (LNA), mixer, power amplifier, and voltage-controlled oscillator (VCO). Most performance parameters for RF circuits can be enhanced mostly by optimizing open-loop parameters, but system-level DC parameters such as offset, image, and spurious tone can be self-trimmed. The bottleneck in RF system designs is the mixer spurious-free dynamic range (SFDR) performance. RF systems can be configured using global feedback and IF quantization concepts, which facilitate the integration of on-chip wireless systems.

6.1 Mixer

The most critical element in wireless systems is the mixer which performs the basic function of frequency translation. It handles nonlinear large signals, and its performance heavily depends on the transient behavior of the nonlinear element. A similar nonlinear functional block found in electronics is a latch, which regenerates the digital output from a small seed signal.

There are two types of mixers performing down-conversion as shown in Fig. 6.1. One is the current-switching Gilbert-type mixer derived from the bipolar technol ogy as shown on the left, and the other is the voltage-switching mixer for the MOS technology on the right side. They are dual in the circuit concept. The former has a high-impedance load for current switching while the latter has a low-impedance load for voltage switching. The tail current is left out on the left since for low-voltage CMOS, the supply voltage is limited. Therefore, the mixer operates

B.-S. Song, *System-level Techniques for Analog Performance Enhancement*,
DOI 10.1007/978-3-319-27921-3_6

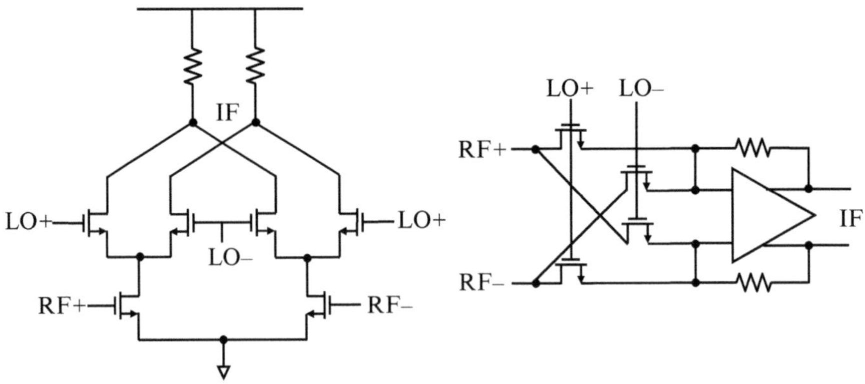

Fig. 6.1 Two down-conversion mixer examples

in a difference mode than a differential mode to save the voltage drop across the tail current source. Although the CM rejection is sacrificed in the difference mode, the RF input to the mixer is converted into current, and the current is modulated by the LO carrier. The modulated current is dumped on the output node with a single pole. The voltage-to-current (*V-I*) conversion as well as the modulation should be linear as they affect the static and dynamic nonlinearities of the mixer. On the other hand, the latter voltage mixer doesn't use the *V-I* conversion, but directly modulates the current before it is dumped onto the low-impedance node given by the transresistance amplifier [1].

The problem with this linear mixing is that the multiplier output gets too small if two small signals are mixed, and the transfer function of the mixer is not linear. Even the mixer offset imbalances the output. That is, the mixer gain is so low that the later stage noise is referred to the mixer input after amplified. To overcome this, mixers for RF receivers are predominantly designed with large local carriers that are large enough to turn the mixer completely on and off like switches. If the chopping or switching waveform is an ideal square-wave, this chopper or switching mixer has an ideal gain of $4/\pi$ independently of the local carrier magnitudes. Since the symmetric square-wave has the harmonics only at odd multiples of the carrier, the chopped output spectrum moves to DC and even multiples of the local frequency. It is a nonlinear transient circuit, and the dynamic transient distortion depends on how the voltage changes from one level to another like the 1b DAC output as extensively discussed in Chap. 1.

During the nonlinear transient period, both noise and distortion peak. When MOS switches are loosely turned on, the mixer noise gets high. The mixer performance is entirely dependent on the static nonlinearity of the *V-I* conversion and the dynamic nonlinearity of the output transient. Both noise and linearity performance can be only reduced by careful design and fast switching.

The mixer shown on Fig. 6.2 combines the features of two mixers shown in Fig. 6.1. The bias current is added to the *V-I* stage as marked by dotted lines, and only the signal current is switched to the load [2]. This mixing requires fast

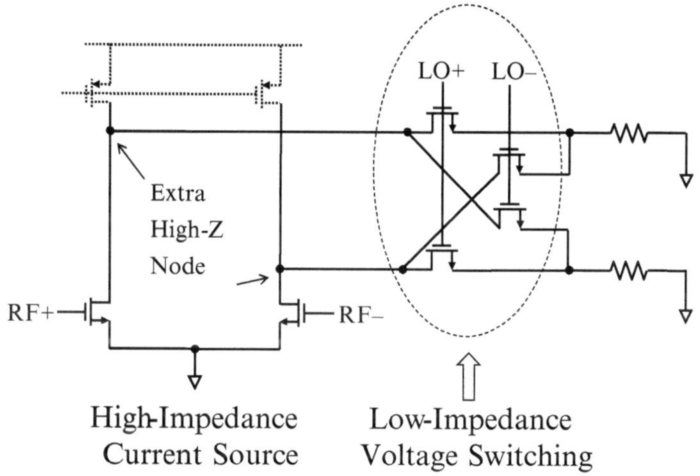

Fig. 6.2 Mixer switching bias current only

Fig. 6.3 Passive and sampling mixers

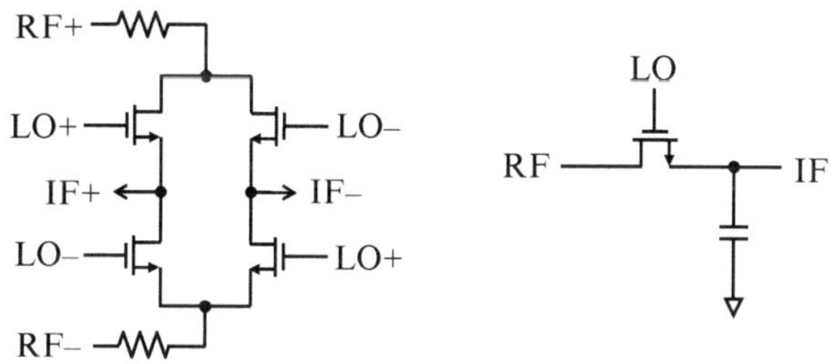

switching since the bias current to the *V-I* block adds extra high-impedance node, which complicates the output transient behavior, and the large bias current source adds noise. Furthermore, the voltage switches do not perform well with high-impedance terminations on both sides. Although only bipolar signal current is switched in the class-B mode, this voltage-switching mixer tends to get nonlinear unless switched fast.

Similar voltage-switching mixer has been named as passive and sampling mixers as shown in Fig. 6.3. Passive sample-hold has also been used as a mixer as sampling can extract the low beat frequency. Very large local carrier is necessary to operate MOS switches. In particular, passive mixers suffer from poor isolation, signal loss, noise, and nonlinearity, and sampling mixers exhibit different sets of problems such as aliasing, jitter, noise, and nonlinearity.

The best performing mixer is the current-switching chopper mixer as shown in Fig. 6.4, replicating the simple single-pole RC load driven by the square wave. That is, it can be approximated as a linear steady-state circuit. The unipolar current is switched to the linear time-invariant load. The RF input is converted into the current in the tail current source, which should have a large overdrive voltage for

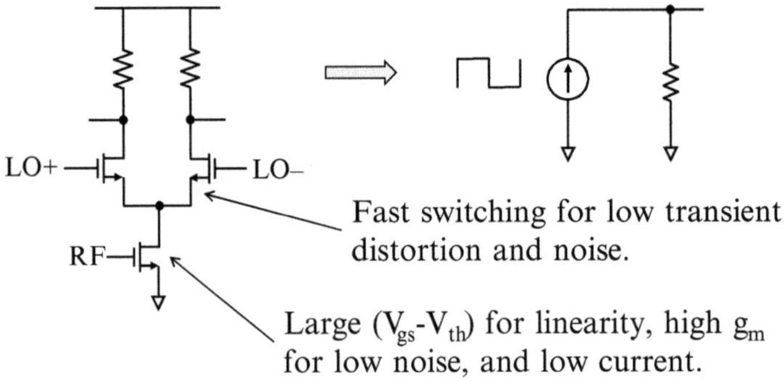

Fig. 6.4 Two-quadrant current-switching mixer

	2-Quadrant	4-Quadrant	Continuous	Switching
Receiver	X	X	Noise	X
Transmitter	Carrier	X	X	X
Output Filter	-	-	-	X
No Filter	-	-	X	-

Fig. 6.5 Choosing mixers

high linearity, and its transconductance should be large for low noise. Therefore, once the process technology is known, the device size and bias current can be set to meet the blocker condition. The output noise also depends on the bias current, and the finite rising and falling times limit both transient noise and distortion. The noise and distortion performance can be improved by shortening the rise-and-fall times. Using current scaled MOS technologies with 1.8 V or lower supply, the achievable SNR and SFDR using chopper mixer is limited to about −70 to −80 dB for 10 MHz bandwidth with the peak signal of 100–150 mV level, which is also about the dynamic range required to meet the blocker specification of most wireless receivers for Bluetooth and WiFi. Therefore, unless AGC is not applied in the baseband after the mixer, the mixer dynamic range sets the sensitivity and the blocker conditions of most RF wireless systems.

Different types of mixers can be chosen depending on their uses as shown in Fig. 6.5, where X marks the safe operational configuration. Two-quadrant mixers are acceptable for receivers, but only four-quadrant mixers work for transmitter since carriers can be suppressed. For receivers, continuous-time mixers are not appropriate due to their high noise levels though they require no output filters. Therefore, switching or chopper mixers with output filters are common. However, switching harmonic mixers don't need output filters.

6.2 Sensitivity and Blocker

As scaled CMOS technology dominates the wireless transceiver market, the RF design paradigm shifts towards integrated low-cost architectures. In the old single-media era, most receivers for TV, broadcast AM and FM, and radios used tuners with highly selective RF front-ends. As the nearby adjacent or alternate channels are tuned out, the RF front-ends don't need to have wide SFDR. However, as digital communications become ubiquitous, the same frequency bands are shared by many users. It was made possible by digital modulations, which spread the spectrum over the wide range both in frequency and time in the form of spread spectrum, frequency hopping, and OFDM. To selectively receive the desired channel in the heavily populated frequency band, the receiver front end should be immune to nearby large interferences (blockers) while receiving small signals. Therefore, unlike old RF receivers, modern ones for digital communications can be characterized by two parameters—noise and distortion. They can be specified by the input sensitivity or the noise figure and by the blocker condition, respectively. The former defines the minimum receivable signal, and the latter specifies the maximum allowed nearby blockers such as adjacent and alternate channels. In most receivers, these two parameters are limited by the mixer.

Traditionally, electronics started from the discrete RF designs as in the bottom-up approach as shown in Fig. 6.6. RF systems are configured using standard individual components that already exist. However, new digital receiver design should start from the top while optimizing individual functional blocks for the system performance. That is, in old RF designs, there are so many parameters to consider, but in new RF designs, only two parameters count as shown in Fig. 6.7.

The old RF design is mathematically based on the linear extrapolation of the receiver transfer function. Since most functional blocks are very nonlinear and have

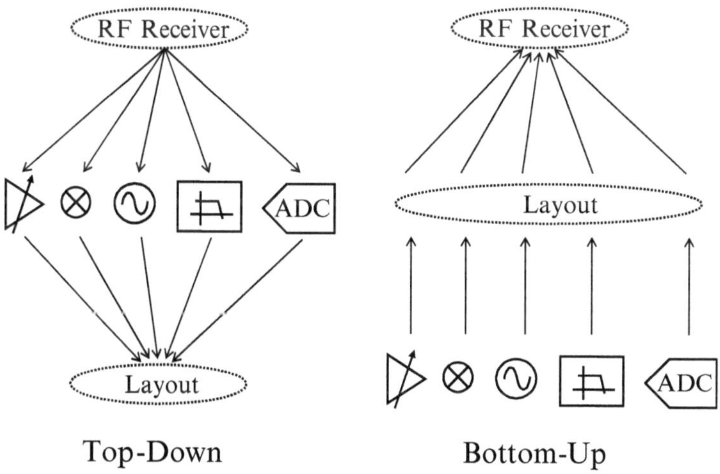

Fig. 6.6 Top-down and bottom-up approaches in RF receiver design

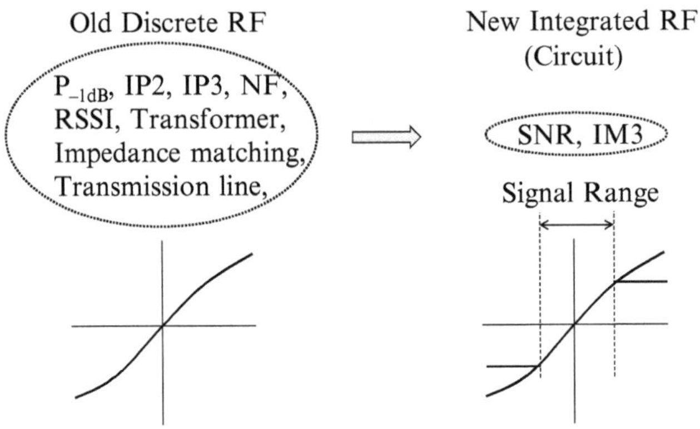

Fig. 6.7 RF design paradigm shift

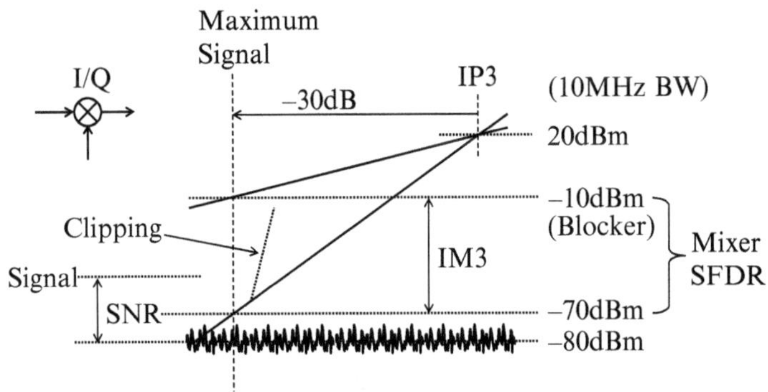

Fig. 6.8 Mixer SNR and SNDR definitions

limited dynamic range, the AGC function is distributed throughout the receiver chain to warrant the linear operation of all blocks. However, the fundamental feature the modern receiver has is the linear absolute signal range that meets both SNR and IM3. Therefore, there is no need for distributive AGC. The global AGC only before the mixer stage will suffice. For that, the mixer SNR and SFDR should be the critical design parameters, on which the receiver performance hinges. All modern digital receivers adopt direct or low-IF down-conversion architectures, requiring two in-phase and quadrature (*I*/*Q*) mixers.

Figure 6.8 sketches the signal and the third harmonic HD3 vs. the input signal. Both are extrapolated to meet at the IP3 point. For the typical 10 MHz BW, IP3 is about 15–20 dBm. The maximum input to the mixer should be limited to −30 dB point from the IP3 point. It implies that at this maximum input condition, the blocker is at −10 dBm, and its IM3 will be at −70 dBm, which should be higher than the noise floor at −80 dBm. The noise of the nonlinear mixer is not clearly defined and depends heavily on the rise/fall times of the local carrier, but

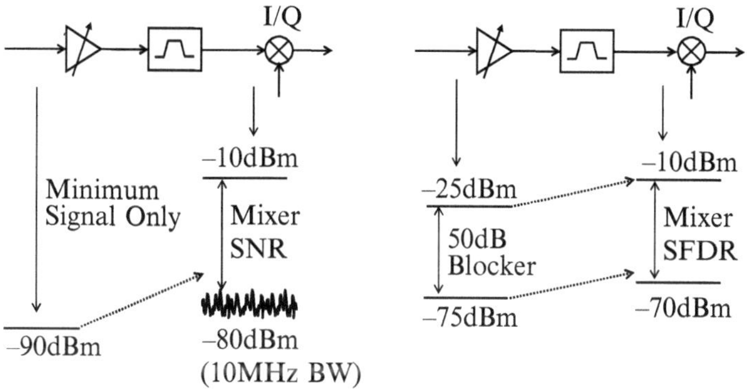

Fig. 6.9 Receiver SNR and blocker conditions

the −80 dBm order noise is typical for the carefully designed chopper mixer. Beyond this point, HD3 increases rapidly due to the clipping of the mixer input stage. Assume that the input signal is amplified or attenuated so that the input LNA noise can be higher than the mixer noise. Then, the signal fits comfortably inside the mixer SNR and SFDR as shown in Fig. 6.9.

Two RF receiver specifications define the minimum signal and the maximum blocker conditions in two different terms. The former is specified with just the minimum signal only often limited by the thermal noise at the RF input port. There is no blocker present at the sensitivity point. On the other hand, the latter blocker is specified for the maximum signal at much higher than the sensitivity point for example at −75 dBm with a 50 dB stronger blocker at −25 dBm. That is, the receiver should have a minimum SNR of over 14 dB for BPSK and higher for other modulations at the sensitivity point and in the blocker condition.

6.3 Global AGC Feedback

As discussed, the mixer DR is the bottle neck in the RF receiver design. The mixer noise and IM3 directly set the system NF and the blocker condition. If the SFDR of the mixer is limited, the system requires further RF filtering of the blocker before the mixer. Therefore, the ADC can quantize the fixed mixer output directly. The overall system with the global AGC is shown with three different ADCs in Fig. 6.10.

Assume that the input range from −90 to 10 dBm is fitted into the mixer SNR using AGC as shown. Most open-loop chopper mixer DR is limited to about 70 dB with the maximum input of 100–150 mV. Therefore, its output can be directly quantized using a 12b CT DSM. If filtered using a fifth-order AA LPF and amplified by 20 dB, a Volt-level Nyquist-rate 12b ADC can be used. If further filtered with an analog channel filter, a 6b ADC suffice. It is true that conventional analog signal

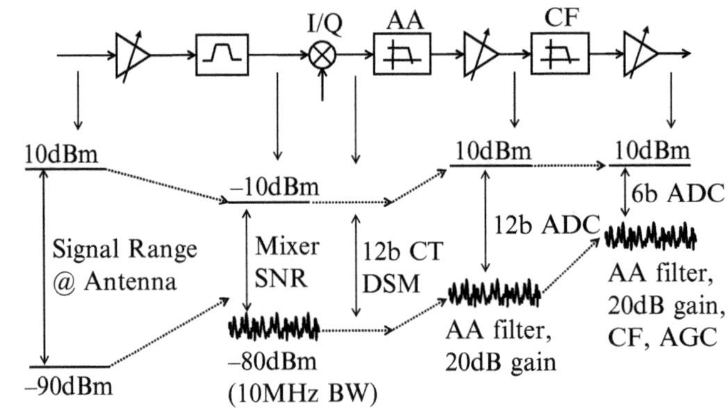

Fig. 6.10 Quantization schemes for RF receivers

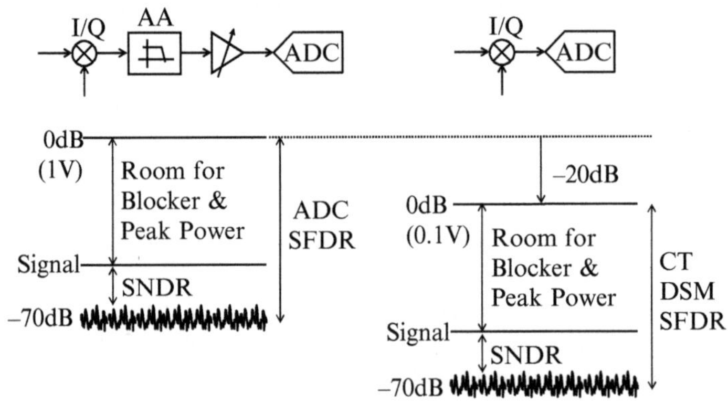

Fig. 6.11 Comparison between Nyquist-rate ADC and CT DSM

processing simplifies the ADC design. However, the trend of moving the ADC closer to the RF greatly simplifies the RF receiver since now all AA and CF are done digitally while quantizing the −20 dB smaller signal.

More than anything else, the global feedback completely eliminates the local AGC after the mixer stage, thereby carrying out the signal level detection for AGC feedback more accurately in the digital domain [3]. For baseband quantization, the oversampling DSM is an ideal choice since it simplifies the AA filtering. Furthermore, the CT DSM eliminates the need for AA filtering entirely, and can even perform some of the blocker filtering if the signal transfer function is modified with zeros at the blocker channels.

Two candidate ADCs, Nyquist-rate and CT DSM, are compared as quantizers in the RF receivers in Fig. 6.11. The CT DSM exhibits a −20 dB lower noise floor, and requires no AA filter. The future trend is towards all-digital AA, CF, blocker filtering, and even AGC. Since all AA and CF are done digitally, it is an ideal ADC for RF receivers in the software-definable radio environment.

6.4 Impedance Matching

RF signal goes through antenna to be received or get radiated. When receiving, the retarded potential field yields the electric field on antenna, and the field induced current through the antenna is collected if loaded by the matched impedance given by LNA. When transmitting, the drive current to antenna from the PA creates the electric field distribution, and the retarded potential field is radiated from antenna. Antenna transfer functions while receiving and transmitting are reciprocal. That is, the electrical field creates the antenna current, and the antenna current creates the electrical field.

The maximum field is induced on the half-wavelength dipole antenna, and the simplest monopole antenna is a half-length dipole antenna over ground plane. Antenna is made of a metal wire, transmission line, microstrip line, loop, or waveguide. The monopole antenna is made of a quarter-wavelength long metal. The cross-sectional area and the length of the antenna affect its electrical characteristics with complex impedance with real resistance and imaginary reactance. The impedance of the quarter-wavelength monopole antenna is complex and about 36.5 $+j21.25\ \Omega$ at the quarter-wavelength frequency. At the higher and lower frequencies, the reactance becomes more inductive and capacitive, respectively. Therefore, if the antenna is tuned at the quarter wavelength frequency, it can be considered as a fixed resistive load as shown in Fig. 6.12.

Most antenna resistances are approximately real at the designed frequency, and takes values of 50, 75, and 300 Ω. Most wireless transceivers and test instruments are terminated by a 50 Ω real resistance. For the maximum power transfer through the antenna port, the input and output impedances of any circuits tied to the antenna should be matched to 50 Ω as sketched in Fig. 6.13.

In the RX mode, LNA input provides a matched 50 Ω load for high sensitivity with low NF. In the TX mode, the load of a common-source PA is also 50 Ω.

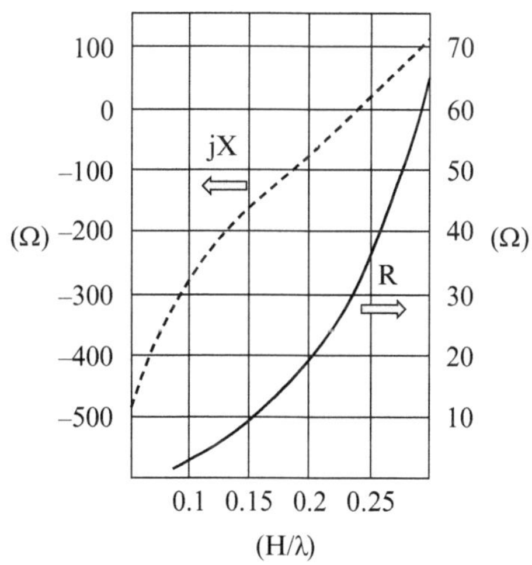

Fig. 6.12 Impedance of quarter-wavelength monopole antenna

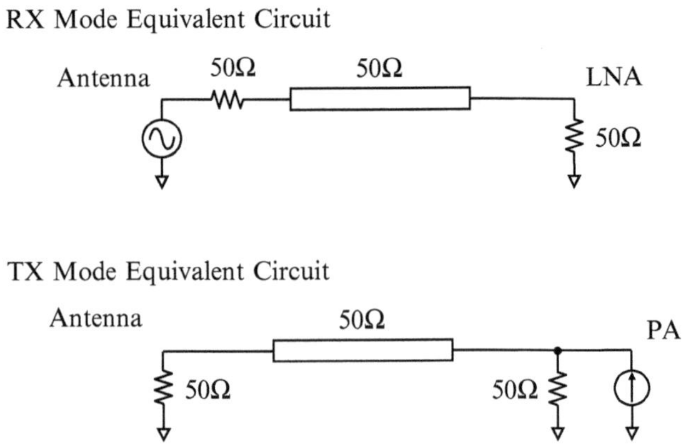

Fig. 6.13 Impedance matching concepts

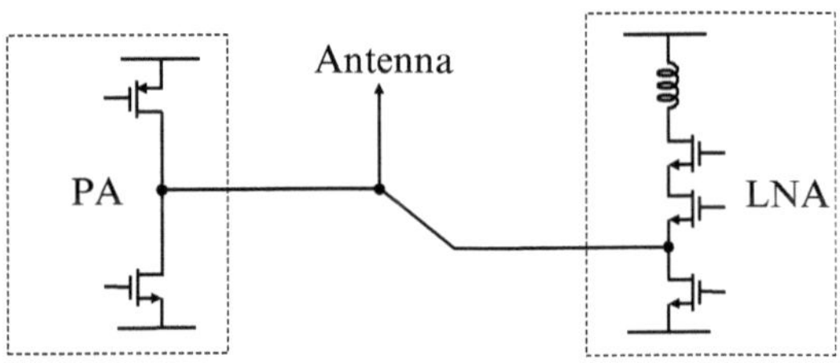

Fig. 6.14 LNA and PA sharing the same antenna port

In reality, the input and output impedances of actual RF chips vary widely with large bonding-wire inductance and parasitic capacitances. Most RF system standards require a VSWR of about 1.5, which allows termination resistances ranging from 33 to 75 Ω. The impedance matching of that level can be achieved without using any elaborate antenna tuning in network standards such as Bluetooth and WiFi. RF switches are commonly used to route the same antenna to RX and TX. However, CMOS RF switches degrade the RF performance since they add noise and parasitic capacitances. Therefore, for simple networking uses, it is convenient to use a common-gate LNA for RX and a common-source PA for TX as shown in Fig. 6.14.

The common-gate configuration makes it possible to share the same RF port with TX and also provides some ESD protection for the RF port [4]. The common-gate LNA exhibits lower NF than the matched common-source LNA, but with Gm-boosting feedback, the NF of the common-gate amplifier can be further lowered. The operation is just to switch on RX and TX parts alternately. The turned-off part will add some parasitic capacitance to the turned-on part. There exists still an impedance matching issue in operating this configuration. In the RX mode, the LNA needs to have a 50 Ω real resistance load for input matching, which is critical in achieving the minimum NF. However, in the TX mode, terminating the TX output with a 50 Ω resistor doubles the power consumption to drive antenna. Therefore, in most TX design sharing the same antenna port, antenna is driven directly with a high-impedance common-source amplifier. It is like a current source

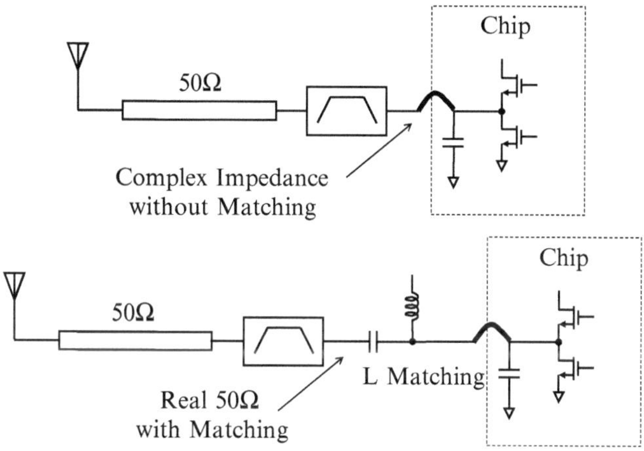

Fig. 6.15 RF port tuning concept

driving a 50 Ω load. This standard common-source PA has no impedance matching while transmitting if the distance from the PA and the antenna is separated farther than the wavelength.

Although antenna tuning is not necessary for most networking applications, high-performance systems such as cellphones require more elaborate input and output terminations, and the impedance matching issue should be considered more seriously. The impedance matching principle is to terminate both the source and load sides of any transmission lines with a real 50 Ω resistance. Any RF ports have parasitic inductors and capacitors from bonding-wire, bonding pad capacitance, package, and device parasitic capacitances. Therefore, impedance becomes complex. In both RX and TX cases, it is mandatory to tune out any parasitic inductance and capacitance using a matching network unless they are negligibly small as shown in Fig. 6.15.

The simplest L-section maintains a matched condition at a fixed frequency, but more elaborate matching networks such as π-section or multi-stage matching networks can be used for broadband impedance matching. Once the complex impedance is tuned out to be real with matching network, the antenna port impedance is considered to be a real 50 Ω resistance at the tuned frequency. That is, the antenna can be connected to the band-pass filter through a 50 Ω transmission line.

Figure 6.16 illustrates the impedance transform to match the PA with the transmission line by stepping up or down the impedance. The connection between the filter and the RF port of the chip should be made short though parasitics at the RF port are tuned out. If the connection between the filter and the chip RF port is a long transmission line, the impedance matching is more complicated in the TX mode. The PA source resistance is higher than 50 Ω since it is a current source. This may cause the situation that both the signal received by antenna and the transmit signal reflected from antenna can be reflected back to antenna due to this impedance mismatch. The PA output can be matched by adding a 50 Ω load, but it consumes

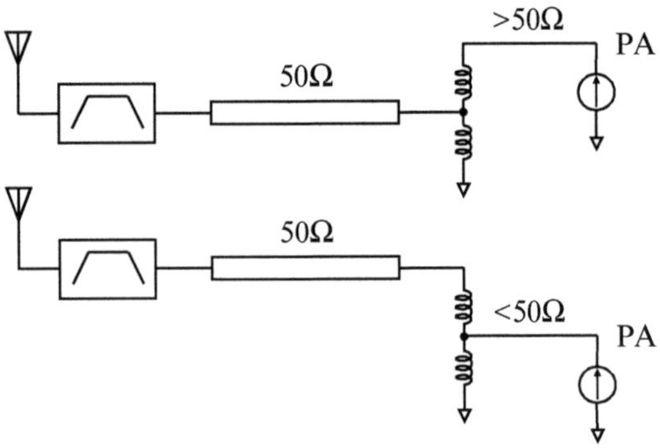

Fig. 6.16 Impedance step-up/-down for matching using a transformer

twice as much power. In such a case, the antenna port cannot be shared. For cellphones, the PA output matching is important to reduce the reflected energy from the transmitter, but for short-range networking applications, it is not considered as serious.

6.5 Digital RF

The most prominent RF design trend is switching digital RF. Discrete-time analog processing is based on two analog functions of sampling and switching. It is basically going back to old analog days with superior digital switching devices. Although some power may be saved in the process of skipping quantization, switching analog signals at RF suffers from both static and dynamic imperfections of analog signal processing. Analog sample/hold or track/hold is nonlinear like the old bucket brigade or CCD due to the switch nonlinearity, charge injection, and clock feedthrough. Only bottom-plate switched-capacitor versions with opamps are practical, but they are slow and noisy due to wideband noise aliasing when sampling.

The fundamental limit of the digital RF is the voltage-to-time conversion accuracy unless the signal exists already in the time domain. The time quantization error is related to the normalized voltage quantization error for the band-limited signal as follows.

$$\frac{(\Delta t)^2}{12} = \frac{1}{12} \times \frac{(\Delta V)^2}{(2\pi \times \mathrm{BW})^2}. \tag{6.1}$$

This implies that to quantize 10 MHz signal with 10b accuracy, it requires approximately a time quantization step of 15.6 ps, which is about the period of 63 GHz. Such Nyquist-rate analog processing is not feasible, and for time-domain signal processing to be practical, the oversampling feedback approach should be considered. However, all other analog problems such as wideband aliased noise, transient

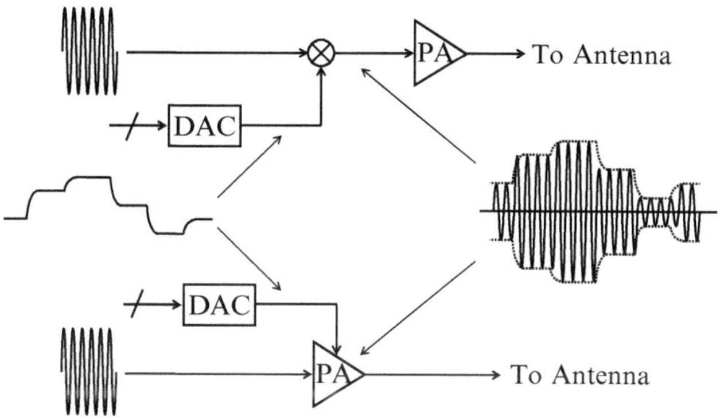

Fig. 6.17 Conventional up-conversion and envelope modulation using RF DAC

nonlinearity, and nonlinearity in the voltage-to-time conversion still remain. Therefore, digital RF processing that applies switching analog signal processing to implement RF transceivers may result in inferior RF performance.

Digital RF has some uses in RF PAs and DACs that require no high SNR or SFDR. Although the linearity requirement is not as severe as for RX, they are large signals, and still the nonlinearity, in particular the dynamic transient nonlinearity remains as the most critical issue. One example of RF DAC and PA is to switch current sources or DAC outputs directly into antenna to generate an envelope-modulated TX signal. When compared to existing RF TXs, which generate a low-frequency baseband envelope only, directly generating the RF carrier requires accuracy with three order of magnitude-faster DAC settling. Two PA examples are compared in Fig. 6.17.

As shown at the top, the envelope of the TX waveform is shaped by mixing, and the linear PA drives the antenna. At the bottom, the PA is replaced by an RF DAC, which shapes the output power envelope directly [5]. The DAC generates an incremental step of the RF envelope. In addition to the switching speed and the step size linearity, there is a risk in switching RF directly to the antenna. The antenna port is not a resistive load but very reactive for the broadband transient signal. Note that linear PAs are mostly for narrowband linear amplification and filtering of steady-state signals. Therefore, if RF circuits are switched, there arises a concern about the spectral regrowth since DAC switching transients are very nonlinear.

There are two reactive nodes between the PA and antenna before and after the band-pass filter as shown in Fig. 6.18. PA injects a step current into this antenna port. The step function has frequency components spanning over a wide frequency range. A problem arises when this broadband transient signal passes through the limited bandwidth stationary channel like a band-pass filter. That is, the filter group delay is a frequency function, and all frequency components experience different delay as well as attenuation. This implies that the step signal loses its integrity, and distorted.

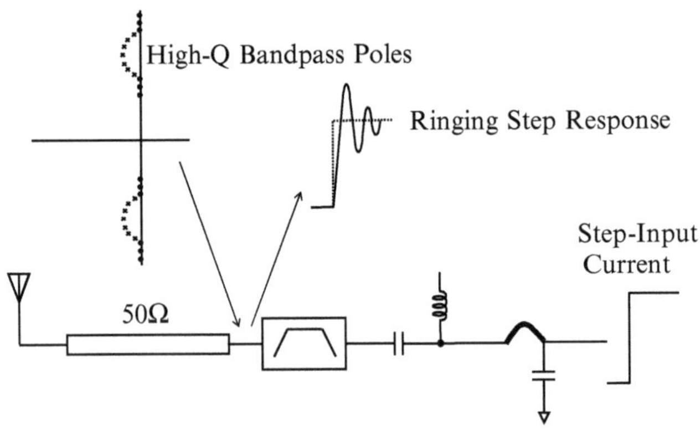

Fig. 6.18 Transient distortion of nonlinear DAC envelope

As discussed in Chap. 1, all individual poles contribute to the transient response. Assuming that the filter input is tuned using an L-section matching network, its impedance is real, but it is impossible to maintain such tuning range over a wide frequency range. Therefore, in practice, it is reasonable to assume the antenna impedance to be complex with reactive components. This implies that its pole should be placed at very high frequencies close to the negative real axis, and the step response should be very fast with ringing. On the other hand, the filter itself has many poles, and its transient response is difficult to predict. Poles are spread over the passband with different Qs. Its step response is expected to be fairly complicated with significant ringing. Therefore, direct switching to antenna is a challenging analog problem that requires sophisticated design skills with superior technology. It may be possible to switch to test instruments matched with broadband 50 Ω terminals. If the transient distortion is poorly controlled, direct switching PA outputs to antenna can cause the spectral regrowth and out-of-band emission.

6.6 Switching Power Amplifier

For high power efficiency, many variations of switching PA have been investigated. It is close to the direct waveform generation at RF frequencies. An extreme example of this digital RF trend is an effort to synthesize the RF binary pulse waveform, and to dump it on the antenna tuning element for filtering. There are two possible versions of RF switching amplifier as shown in Fig. 6.19.

They are class-D and class-B amplifiers which are based on the steady-state filtering principle. If the inputs are voltage and current square-waves, they are filtered by the series and parallel resonant circuits, respectively. The former low-pass filters either the pulse-duty or pulse-density modulated signal to extract the low-frequency signal with high efficiency. It has substantial followings in implementing low-power microphone, earphone, or audio amplifiers. On the

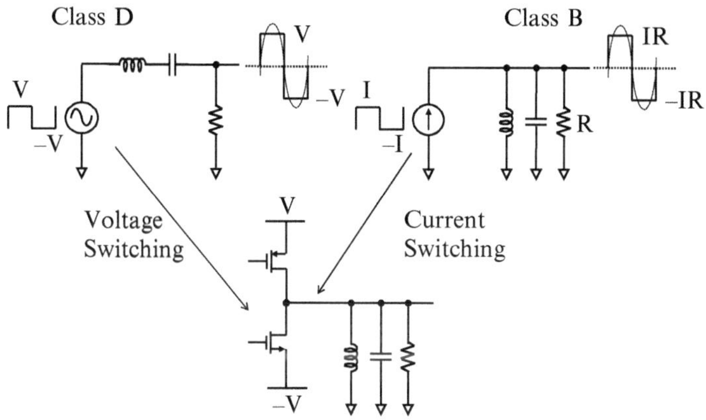

Fig. 6.19 Digital RF amplifier combining two RF switching amplifiers

Fig. 6.20 Switching RF amplifier loaded with transformer

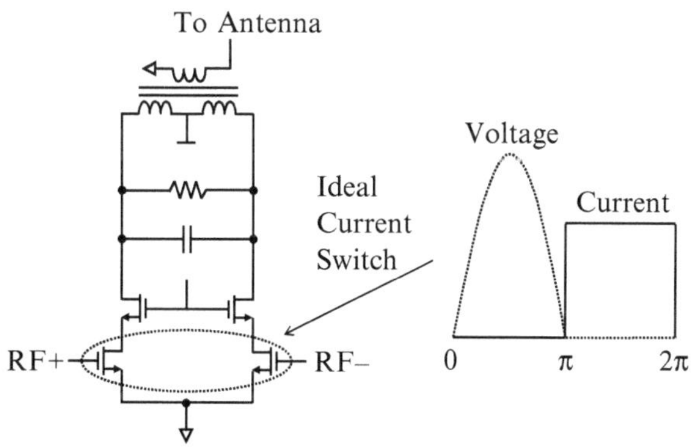

other hand, the latter is a linear amplifier with no standby bias current. The high-impedance tank resonator filters out the resonance frequency. Both are filters, and should be stationary in steady-state.

However, applying the switching principle to the RF amplifier design is a very challenging task as the dynamic transient nonlinearity should be dealt with. The only possible RF switching amplifier takes an intermediate form with neither voltage nor current inputs. The tank circuit is driven by the digital inverter, which is a large-signal nonlinear amplifier mostly slewing or switching back and forth between two supply voltage levels. It is neither class-D nor class-B. It is not a steady-state linear amplifier, but a transient circuit.

Figure 6.20 shows a concept of the digital amplifier even loaded with a transformer [6]. If two switches are ideal with infinitely short rise/fall times, the current through the switch approaches the square-waveform, and the voltage across it should be zero. On the other hand, the resonant tank circuit creates the sinusoidal output if the switch is turned off. All power is delivered to the load without loss to yield high-power efficiency. However, this PA is somewhat flawed in practice.

It suffers from nonideal switches, low-Q resonance, and the lack of transformer operation.

The most critical in the assumption is that it is not the steady-state filtering circuit. The resonance tank with memory is switched like memoryless circuits. That is, the switch should operate only when the voltage across it is zero, so-called zero-value switching (ZVS) not to short the capacitor. Since steady-state filtering requires longer time for filtering, it operates basically like time-domain waveform synthesis as shown in Fig. 6.21.

The partial time responses of the resonator during each clock phase can be pieced together to form a continuous waveform. However, inevitable mismatches between +/− partial responses and clock distort the synthesized waveform. The final value of one transient cycle becomes the initial value of the next one, and the cross-over error or distortion varies with the beat frequency. Tuning or filtering is not possible with time-varying resonator heavily perturbed. As a result, the ZVS is impractical, and the cross-over distortion rises. The signal injection is neither by voltage nor by current. Assume the switch on-resistance is very small like 0.1 Ω. Then, the voltage driving the primary inductance varies like a square-wave, and the transformer load is switched back and forth as shown in Fig. 6.22.

The resonant tank load becomes a time-varying network. It implies that the inductor current I_L is not a square-waveform, but a ramp charging up and down as

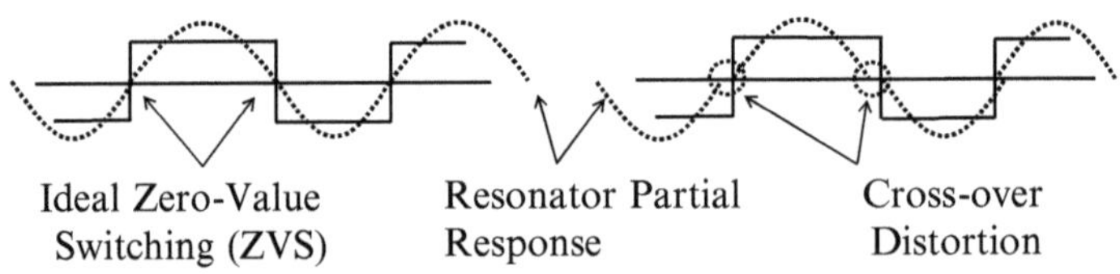

Fig. 6.21 Waveform generation with ZVS

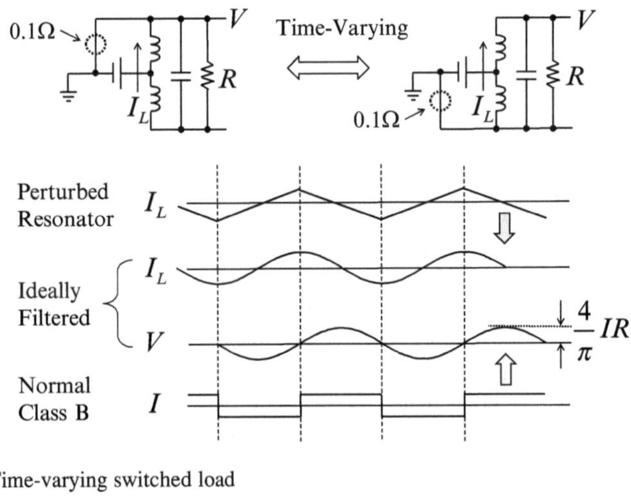

Fig. 6.22 Time-varying switched load

sketched at the top. The resonant partial time response is assumed to be filtered as sketched. However, for the proper operation, the switches should work like a square-wave current source as sketched at the bottom to be in phase with the output voltage. This inverse class-D PA operates in an ambiguous way based on the fact that the inductor is charged both with voltage and current. Injecting current into inductor is power consuming like charging capacitor with voltage, and should be avoided. Although the pulse-width-modulated (PWM) RF binary input is driving the primary inductor directly, the RF transformer attenuates the output significantly.

Transformer is a circuit element as shown in Fig. 6.23 with a 2:1 ratio. If the inductor length is much smaller than the wavelength, a uniform current flowing in the one winding of the transformer creates a flux field, and the current is induced in the other winding by the flux coupling like mutual inductance. The total flux is defined as the product of the inductance L and the current I. If one winding is made twice as long as the other, the inductance is doubled, but the current drops by half. The same ratio can be achieved by increasing the cross-sectional area to get the low inductance with the same length. Note that the flux coupling induces the current in the same direction both in the primary and the secondary sides. Tapping the inductance or the capacitance of the resonant tank in the middle gives the ratio as in the transformer. They all offer the voltage, current, and impedance scaling effects.

As frequency goes higher, even one circuit node (one node in circuits) is spread like the transmission line, and the circuit elements are implemented differently as shown in Fig. 6.24. An inductor can be implemented using a transmission line

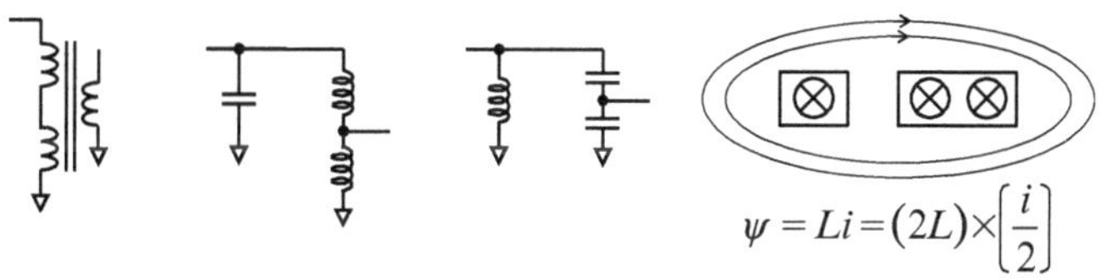

Fig. 6.23 Examples of 2:1 transformer

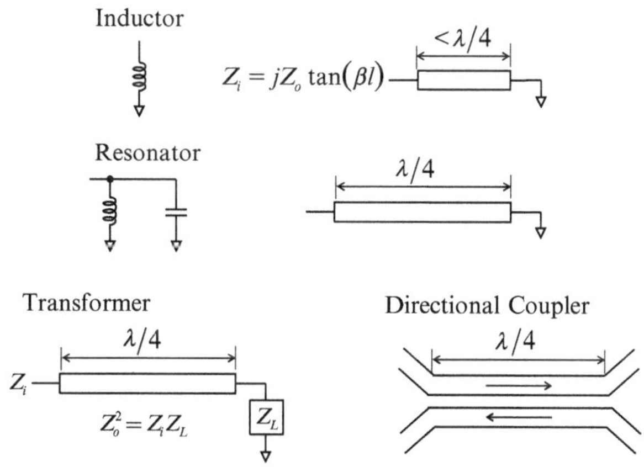

Fig. 6.24 Circuit elements in transmission lines

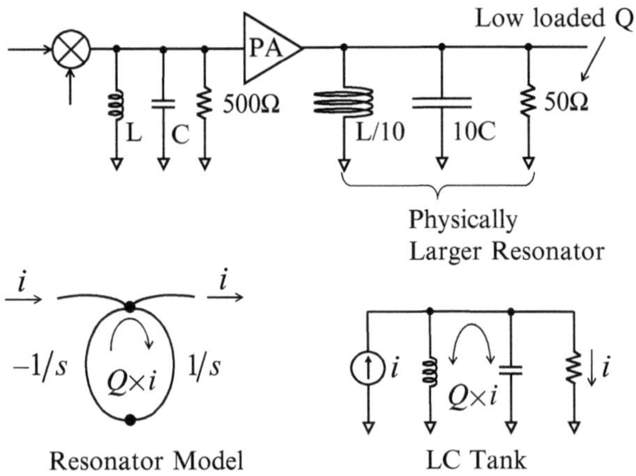

Fig. 6.25 Loaded-Q of RF resonant tanks

which is much shorter than the quarter wavelength ($\lambda/4$). The inductance is a tangent function of the electrical length βl, where β is the wave number and l is the line length. If the length l exceeds $\lambda/4$, it becomes a capacitor. Similarly an LC resonant circuit can be made using a shorted $\lambda/4$ transmission line.

The $\lambda/4$ transmission line is very useful to make a transformer or impedance transformer as shown. Also a two-port directional coupler can be made by coupling two $\lambda/4$ transmission lines, but note that the direction of the wave propagation is reversed. That is the difference between the two-port transformer in circuits and the two-port directional coupler in transmission lines. Therefore, cares must be taken in realizing an RF transformer for PA [7]. The transformer in the circuit is a magnetic flux coupled device while the transmission-line transformer or directional coupler is a capacitive fringe field coupling device. One more constraint imposed on the RF transformer is its bulkiness as explained in Fig. 6.25.

The size of the resonant tank circuit matters most. For example, at the mixer output before the PA, the LC tank can be made small since its loading is light with typically a couple hundred Ω's as shown with 500 Ω. However, the PA output should drive a 50 Ω low-impedance antenna. So to achieve the same Q, ten LC tanks should be connected in parallel to drive the 50 Ω. It is because the resonator holds Q-times higher energy inside the tank. The input and output coupling is through the $1/Q$ coupling path. That is, the energy inside the high-Q tank increases or decreases slowly. Since on-chip inductors or transformers are limited in size, it is not realistic to expect on-chip high-Q RF tuning at the PA output port.

6.7 Fractional-N Frequency Synthesizer

RF receivers need frequency synthesizers for frequency translation of received and transmitted signals. With data receivers trying to achieve high bandwidth efficiency, low-spurious fractional-N frequency synthesizers designed for low phase noise are in very high demand. The Digital Video Broadcasting terrestrial

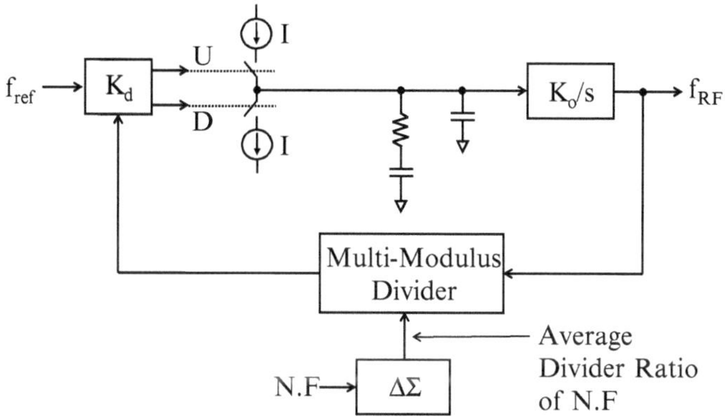

Fig. 6.26 Fractional-N frequency synthesizer based on CP PLL

transmission (DVB-T) standard uses the orthogonal frequency digital modulation (OFDM) with data-rate channels separated by 1 kHz spacing, and needs very low spot phase noise to avoid the degradation in BER due to the adjacent channel noise. The US Advanced Television Standards Committee digital cable (ATSC-C) and DVB-C standards use the 256 quadrature amplitude modulation (256-QAM), which requires very low-integrated phase noise. Symbols on the outer periphery of the constellation suffer the most because of the phase noise. For example, to achieve a modulation error rate (MER) greater than 30 dB for J.83/B, the integrated phase noise should be lower than 1°.

A charge-pumped (CP) fractional-N PLL synthesizer as shown in Fig. 6.26 uses a ΔΣ divider ratio modulator, and achieves low-phase noise by widening its loop bandwidth. The phase noise transfer function (PNTF) of the VCO is similar to the quantization NTF of the DSM. The inverse of the open-loop gain high-pass shapes the VCO phase noise within the PLL loop bandwidth. The widened loop bandwidth results in the substantial suppression of the in-band VCO phase noise, which normally dominates the PLL phase noise.

Also wide PLL loop bandwidth is desirable in some transmitters, which use a direct in-loop data modulation of an RF carrier [8]. Usually this in-loop modulation is preceded by a filter, which compensates for the PLL loop response to enable wideband data modulation. If a wideband PLL is used, this pre-compensation filter is no longer needed. The relation between the PLL bandwidth and the phase noise is shown in Fig. 6.27.

The total integrated phase noise includes the contribution of VCO, ΔΣ ratio modulator, charge pump, and reference noise. Integer-N synthesizer suffers mainly from the VCO noise due to its common narrowband design. By widening the PLL loop bandwidth, a lower integrated phase noise can be achieved until it is limited by the charge pump and reference noise. However, wide bandwidth would mean high-reference frequency, which in turn results in low-frequency resolution. On the other hand, wideband fractional-N synthesizer circumvents this trade-off, and achieves high-frequency resolution, giving greater design flexibility at the system level.

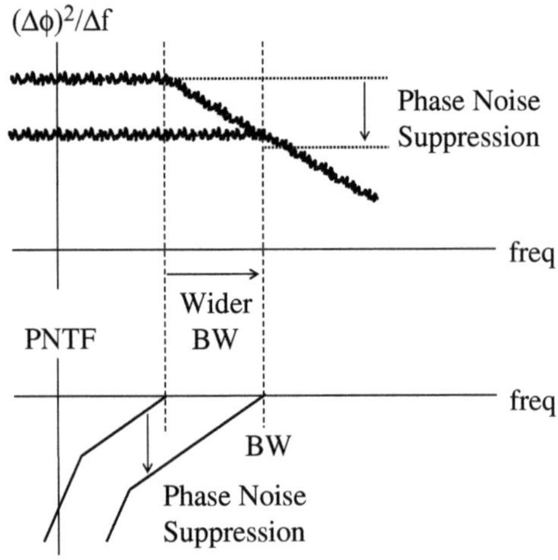

Fig. 6.27 In-band phase noise spectral density vs. loop bandwidth

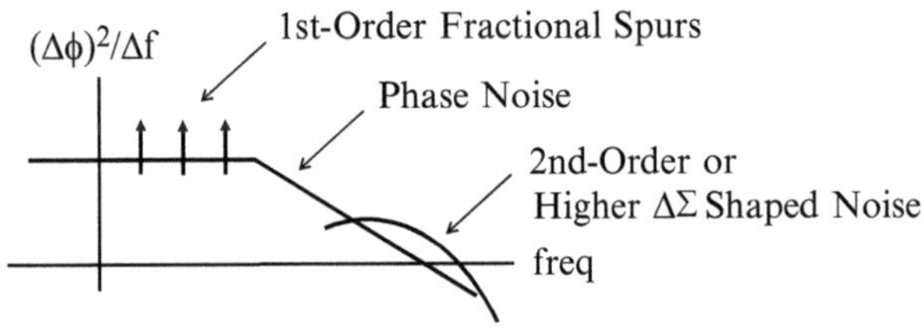

Fig. 6.28 Fractional spurs and shaped noise

6.7.1 Fractional Spur

In fractional-N synthesizers, the frequency divider averages many integer divider cycles over time to make an effective fractional divider ratio. A simple accumulator with an overflow can be used to control an $N/N+1$ divider to interpolate any fractional values between N and $N+1$ as in the first-order DSM, but generates strong periodic fixed tones.

Figure 6.28 shows the effect of the fixed tones and the high-pass-shaped noise added to the phase noise spectrum. To avoid the periodic tones, higher-order $\Delta\Sigma$ divider ratio modulators have been used [9–15]. However, representing a fractional number with a series of integer numbers still produces high-pass-shaped ratio noise similar to the DSM quantization noise, which affects the phase noise performance significantly.

If the loop bandwidth is made wide, the high-pass-shaped noise becomes dominant, and the total integrated phase noise increases. To sufficiently attenuate this high-frequency noise, the loop filter should be narrowed. This bandwidth constraint reduces the PLL loop gain that is needed to suppress the VCO phase noise. This translates into the similar noise and bandwidth trade-off as in the

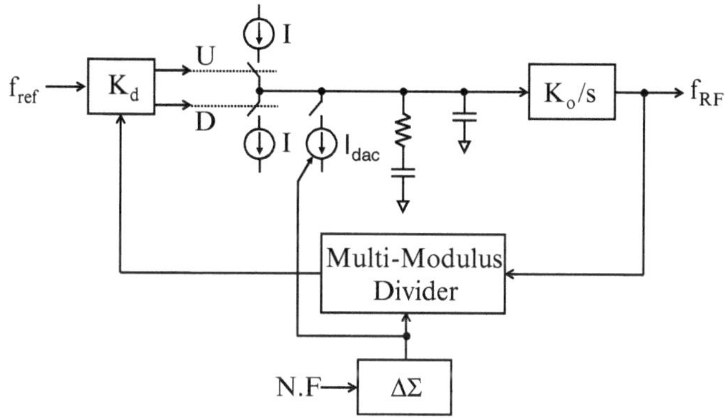

Fig. 6.29 Spur-cancelled fractional-N synthesizer using DAC

integer-N synthesizer case. That is, in the integer-N PLL, the frequency resolution limits the loop bandwidth, but in the fractional-N PLL, the high-pass-shaped divider ratio noise limits the loop bandwidth. Spur-cancellation techniques have been suggested to suppress both the fixed tones and the high-pass-shaped noise below the integer-N PLL phase noise while keeping the PLL loop bandwidth wide.

A DAC-based spur-cancellation concept called digi-phase has been applied to an accumulator-type first-order fractional-N PLL [16], and later generalized for higher-order ΔΣ fractional-N PLLs as shown in Fig. 6.29 [17–21]. The idea is to subtract the dominant divider ratio error in real time using a DAC. If the ΔΣ divider ratio noise is cancelled accurately, the phase noise performance of the fractional-N PLL can approach that of the integer-N PLL, irrespective of the order of the DSM used. Since there is no need for the noise and bandwidth trade-off, it is possible to widen the PLL loop bandwidth further independently of the frequency resolution and the high-pass-shaped divider ratio noise.

Spur averaging also achieves the same spur cancellation. It relies on the fact that the ratio error in a first-order ΔΣ PLL is periodic, and in a time period corresponding to the fractional frequency, the total ratio error charge injected by the CP is zero. Charge errors are accumulated on a capacitor for the time period, and dumped on the loop filter at the end of each period [22, 23]. But these methods are limited in frequency resolution and loop bandwidth.

Many nonideal factors such as phase frequency detector (PFD) and CP nonlinearity, DAC resolution, ratio-dependent jitter, and DAC/CP gain matching limit the effect of spur cancellation, and leave residual spurious tones in the PLL output. The PFD/CP nonlinearity can be avoided by operating the CP away from the nonlinear zero-crossing point [18]. Furthermore, the DAC resolution can be improved by shuffling, and the ratio-dependent jitter can be removed by re-latching the feedback clock. However, the critical DAC/CP gain mismatch problem has not been addressed clearly though it has been identified as the most important limitation of the spur-cancellation techniques [16, 18–20], and the effectiveness of the DAC-based spur cancellation critically depends on it.

A sign–sign LMS zero-forcing feedback concept can correlate the actual accumulated divider ratio error at the loop filter with its sign polarity, and the DAC gain is adaptively trimmed based on the sign of the detected gain mismatch error so that the DAC/CP gains can be closely matched.

6.7.2 DAC-Based Spur Cancellation

A fractional-N divider ratio modulator takes a fractional word *N.F*, and generates a sequence of time-modulated integers, whose time average is *F*. The VCO frequency is divided by the time-modulated divisor values. If locked, the RF VCO frequency f_{RF} is equal to $(N.F)f_{ref}$, where f_{ref} is the reference frequency. Then, the divider and reference edges differ by a fraction of T_{VCO} as shown in Fig. 6.30.

This ratio error is accumulated, and the phase error at the PFD output is fed back. Spur-cancelled frequency synthesizers calculate the value of the cancellation charge, which estimates a PWM charge error. It has a fixed magnitude but a variable width depending on the divider ratio. To cancel this CP pulse, a PWM cancellation charge is necessary. However, generating such a cancellation charge requires a very precise time slicing on the order of ps. To avoid this, a pulse-amplitude modulated (PAM) DAC cancellation charge with a fixed pulse-width can be used as shown in Figs. 6.31 and 6.32.

The PFD/CP-related nonlinearity is one of the well-known sources of spurs, which are more prominent at the integer boundaries of the fractional words. It primarily results from the PFD dead zone and the CP UP/DOWN current mismatch, and also from the rise/fall time mismatch. In practice, it is a PWM 1b DAC, and its transient distortion is fundamental in all analog designs. It creates spurious tones in the PLL output. If only the static current gain mismatch between the UP/DOWN pulses is considered, harmonics and subharmonics (due to spectral folding) are generated at nFf_{ref}, where *n* is the harmonic number. To reduce the PFD/CP related nonlinearity, the CP can be operated in a class-A mode away from the nonlinear zero-crossing point. For that, a fixed timing offset is introduced in the CP

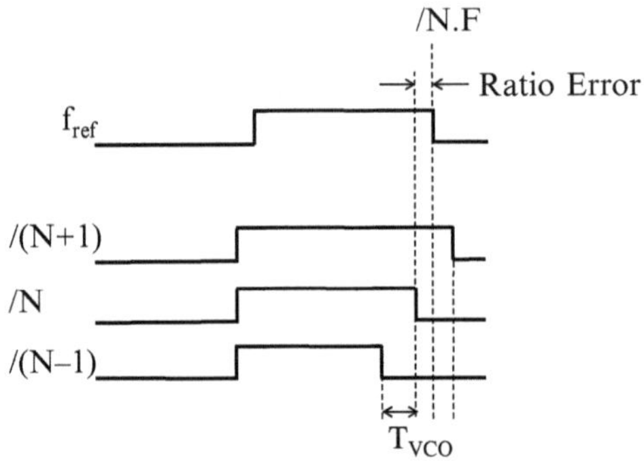

Fig. 6.30 Phase error from the fractional-N divider

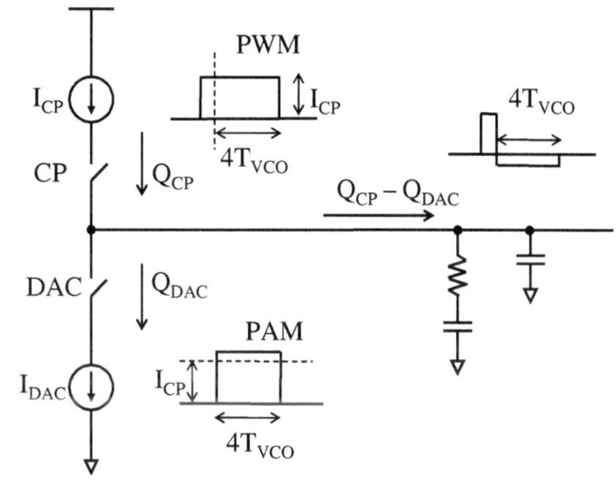

Fig. 6.31 Cancelling PWM error with PAM DAC

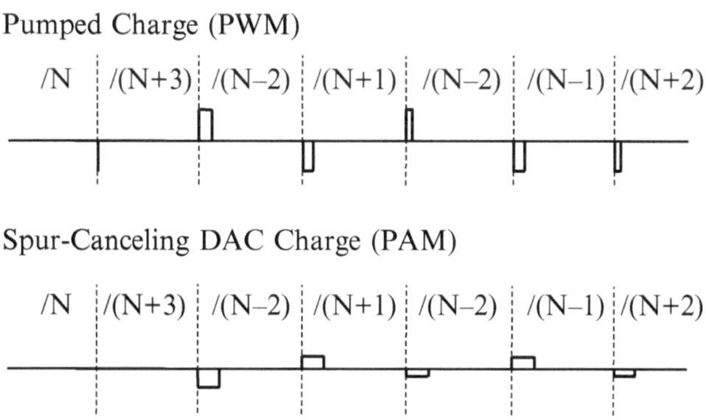

Fig. 6.32 PWM charge error vs. PAM cancelling charge

UP/DOWN pulses by sinking a fixed offset current for a time period so that only the UP pulse modulation can be used to maintain the steady locked state. But in the transient state, both UP/DOWN pulses can be active. Also the rise/fall time mismatch-related transient nonlinearity is reduced because the timing offset is much longer than the rise/fall time of UP/DOWN pulses. However, the pulse-width cannot be very long as it will increase the CP-phase noise. A shorter pulse-width will lead to more rise-and-fall-related nonlinearity and jitter contribution to the phase noise.

This DAC/CP gain mismatch results in the residual charge error at the loop filter. This puts a constraint on how wide the PLL loop bandwidth can be and limits the minimum achievable phase noise. The DAC/CP gain mismatch not only results from the DAC/CP current mismatch but also from the rise/fall time difference of the DAC/CP current pulses. That is, the feedback error due to the intrinsic noise sources (VCO, CP, etc.) is not cancelled, but the feedback error due to the $\Delta\Sigma$ ratio-divider noise decreases with DAC/CP matching. The DC value in the feedback error is cancelled by the PLL loop. Although the static current mismatch contributes most to the DAC/CP gain mismatch, the rise/fall time difference of the DAC pulse also contributes to the gain mismatch. The mismatch effect is severe since T_{VCO} is short. With 10 % DAC/CP gain mismatch, it is typically limited to about 0.7°, and the spot noise is at least 10 dB worse. Therefore, in order to get the lowest possible phase noise, the DAC/CP gains need to be accurately matched. The high phase noise of the PLL example is due to the increased ΔΣ noise because the bandwidth of the PLL is widened.

6.7.3 LMS-Based DAC Gain Calibration

Fractional-N synthesizer can be configured to address most of the issues related to the spur cancellation with an emphasis on the DAC/CP gain mismatch. The fractional-N synthesizer with a spur-cancellation DAC is shown in Fig. 6.33 [24, 25].

There are two added blocks, spur-cancellation and spur-correlation blocks. The spur-cancellation block cancels the ΔΣ divider ratio noise, and the spur-correlation block detects the DAC/CP gain mismatch error for LMS adaptation. Together they complete an adaptive zero-forcing feedback based on sign–sign LMS algorithm, which differentiates it from other spur-cancellation works. The algorithm relies on the direct correlation of the actual accumulated ratio error at the loop filter with the sign of the integrated phase error. A 1b oversampling DSM converts this correlation to a 1b digital bit stream, which is then fed into the digital block for low-pass filtering.

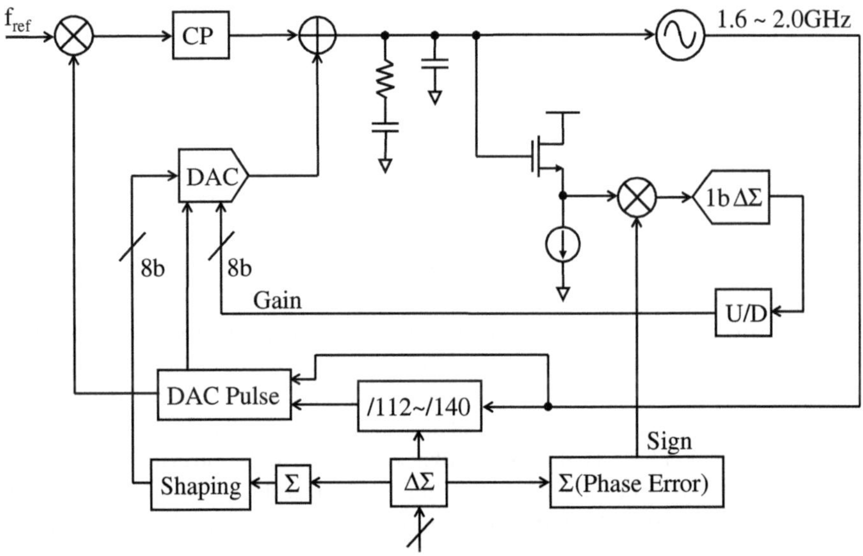

Fig. 6.33 Fractional spur cancellation by LMS adaptation

The word width of the DAC is also an important design factor. For the perfect cancellation of the ΔΣ divider ratio noise, the DAC resolution on the order of the frequency resolution is needed. That is, a 20b DAC is needed for the frequency resolution of $f_{ref}/2^{20} \sim 14$ Hz with a 14.3 MHz reference clock. This 20b DAC is not possible. Using a low-resolution 8b DAC for a 20b frequency resolution creates periodic quantization noise errors, which depend on both the word width and the fractional word. For a DAC resolution of N bits, spurs occur at $f_n = nF2^N f_{ref}$. For the F values close to zero or one (because of spectral folding), the DAC quantization noise is more severe because the harmonics fall in the loop bandwidth. Also the higher the N is, the higher will be the harmonics, and the lower will be the DAC quantization noise power. To alleviate the problem of F related harmonics, the DAC output is processed using a third-order DSM. This modulator takes the full 20b resolution phase error and generates an 8b DAC output. The idea is to high-pass shape any quantization noise generated by the 8b truncation process. The third-order modulator moves this error out-of-the-loop bandwidth where it can be sufficiently filtered by the PLL loop filter.

Although total PWM charge is equal to the total PAM charge, they do not match perfectly in time and frequency domain, and therefore leave some residual voltage on the loop filter for the duration of the PWM and PAM pulses. There is a perfect cancellation of the charge at DC, but the gain mismatch at high frequencies exists. Such voltage perturbations remain as uncancelled ΔΣ divider ratio noise at high frequencies. This gain error can be made negligible if the DAC pulse-width is minimized to reduce the frequency-dependent gain mismatch. This second-order effect becomes important if a much wider PLL bandwidth is desired, and the digital compensation is necessary [19]. The DAC nonlinearity has the same effect as the PFD/CP nonlinearity on the PLL output phase noise. The DAC nonlinearity results mainly from the mismatches among the DAC elements, and can be greatly reduced by using a thermometric structure and/or by using dynamic mismatch shaping techniques [18, 20]. The DAC nonlinearity will also result from the rise/fall time mismatch, and care should be taken to minimize the effect by maximizing the slope of the pulse edges.

An up/down counter controls the gain of the DAC. A second-order DSM generates control word for the frequency divider. The difference between the control word and the 21b fractional word is integrated to get the total phase error of the charge pump. This integrated phase error is a 22b digital value, which is then passed through the third-order DSM that quantizes the phase error to an 8b word and high-pass shapes the resulting quantization noise. The 8b word is converted to 32 thermometer-coded bits and 3 binary-coded bits, which are used to control the switches in the DAC. The strength of this correction algorithm lies in the correction of the DAC/CP gain mismatch based on the actual perturbations occurring at the loop filter due to the gain mismatch using sign–sign LMS correlation. The charge error at the loop filter is correlated with its sign to obtain the correlation energy, which is used to calibrate the gain of the DAC current. Hence the correlation with the sign sequence gives us the power of removing the uncancelled residual ΔΣ divider ratio noise remaining in the loop filter. The sign of its DC value is used to update the gain of the DAC current based on the sign–sign LMS algorithm [26]. Since DAC gain coefficients are updated based on the DC energy of the correlation, any other DC energy affects the minimum resolvable gain error. One main source of the DC offset is the difference between the loop filter voltage and the common mode voltage of the 1b DSM. To further ensure that any remaining DC offset does not affect the update decisions, the sign sequence is made to have a zero mean so that the DC component can be chopped and averaged out. The offset of the DSM can be also calibrated in the digital domain.

The spur-correlation block consists of a source follower buffer, a 1b DSM and two digital accumulators. The key to the proposed spur-cancellation scheme is how to detect the small residual divider ratio error accumulated on the loop filter, and then how to correlate it with its sign sequence without disturbing the loop filter voltage. The source follower buffer provides isolation to the loop filter from the switching transients created by the 1b DSM as shown in Fig. 6.34.

The ADC sampling capacitors are made small to minimize the current transients in the buffer. The sign correlation is performed by switching the input sampling capacitors, which generates non-zero DC value if the DAC and CP gains are mismatched. The 1b output of the DSM is accumulated using a 2's complement 24b accumulator, which effectively low-pass filters the bit sequence. The 2's complement output is sampled using a slower clock than the DSM sampling clock, and subsequently the accumulator is reset. This sampling clock determines the accuracy and speed of the gain calibration. A faster clock speeds up the calibration cycle, but makes the decision error larger. A programmable counter is used to generate a variable slow clock period. With a 20b counter, it takes about 1 s

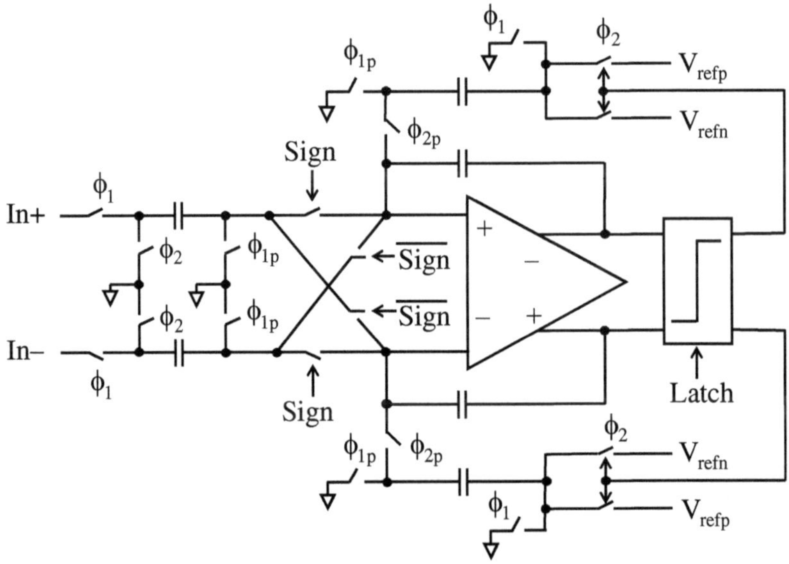

Fig. 6.34 A 1b switched-capacitor DSM

for calibration starting from initial DAC/CP gain mismatch of 10 %. This speed is fast enough to reduce the gain mismatch to below 1 %. The sign polarity of the 2's complement output is used to control an 8b up/down counter, which directly trims the DAC current source bias.

6.7.4 Experimental Results

A 1.8 GHz synthesizer is designed in 0.18 μm CMOS to have a loop bandwidth of 400 kHz and a frequency resolution is 14 Hz with 14.33 MHz clock. The loop filter zero and pole are placed at 100 kHz and 1.6 MHz, respectively. The CP current is 900 μA, and the nominal VCO gain is 40 MHz/V. To attenuate the high-frequency noise more, the PLL loop bandwidth is usually set to be narrow in most fractional-N designs [24, 25]. In the wide bandwidth example with a low reference clock, the loop filter alone cannot fully suppress the high-pass-shaped ratio noise.

Figure 6.35 compares the phase noise spectral densities of integer-N synthesizer and fractional-N synthesizer with a second-order ΔΣ modulus divider. Two phase noises of the fractional-N synthesizer are overlapped to show the difference the spur cancellation makes. Once cancelled, it approaches that of the integer-N synthesizer as shown. If the first-order ΔΣ modulus divider is used, the spur cancellation suppresses fixed tones as shown in Fig. 6.36.

Figure 6.37 shows three measured phase noise plots before DAC correction, after DAC correction but before LMS gain calibration, and after DAC correction and LMS gain calibration. The in-band phase noise of the integer-N synthesizer is −101 dBc/Hz, and the spot phase noise is −118 dBc @ 1 MHz offset. The integrated phase noise from 1 kHz to 10 MHz is 0.68°. In the fractional-N mode before the DAC correction, the integrated phase noise is measured to be high 6.37°. The high phase noise has resulted from the high-pass-shaped divider ratio error of the second-order ΔΣ divider ratio modulator used in this work. A third-order DSM found in most fractional-N synthesizers would result in a slightly lower integrated

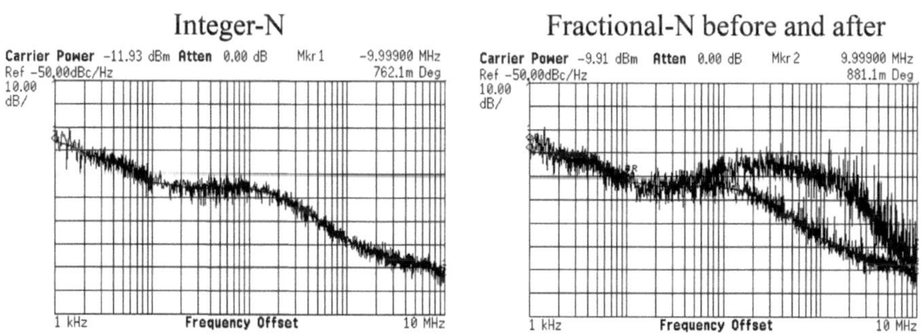

Fig. 6.35 Fractional-N phase noise before and after spur cancellation

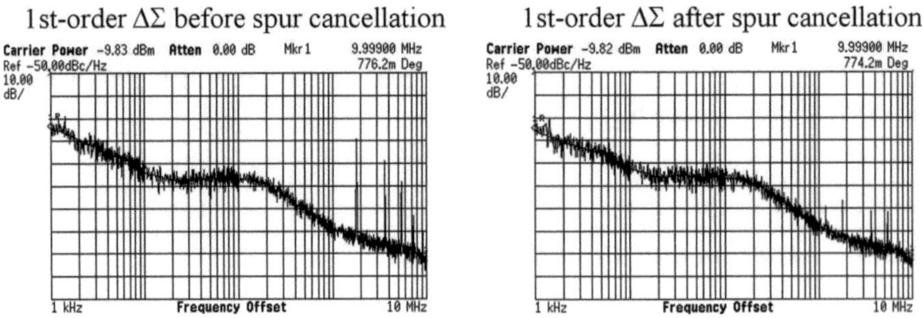

Fig. 6.36 First-order fractional-N phase noise before and after spur cancellation

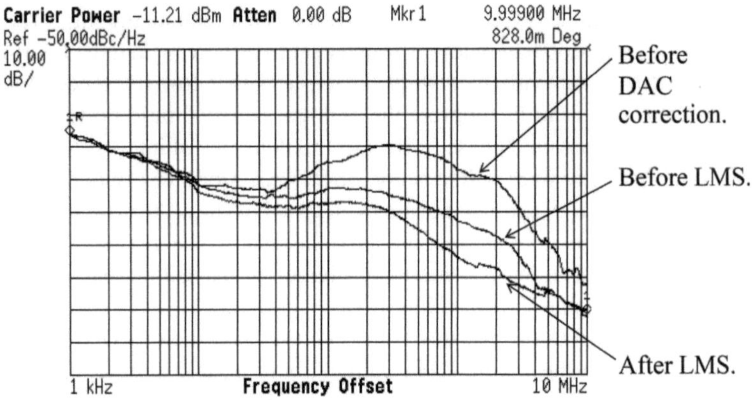

Fig. 6.37 Fractional-N before DAC correction and LMS

phase noise due to its steeper high-pass noise shaping, but the end result will be about the same after spur cancellation. After the DAC correction, the total integrated phase noise is reduced to 1.28°. The LMS-based DAC/CP gain mismatch calibration further reduces the in-band phase noise and the total integrated noise to −98 dBc/Hz and 0.82°, respectively, which approach the phase noise performance of the integer-N synthesizer within a close range of 3 dB and 0.14°.

Figure 6.38 shows the corresponding wideband spectra of three-phase noise curves. This measurement demonstrates how severely the DAC/CP gain mismatch affects the phase noise performance. The spot noise is at least 10 dB worse if the gain mismatch is not corrected. The total spot noise reduction of as much as 30 dB is achieved by the LMS-based gain mismatch calibration. The fractional tones are well below the −70 dB noise floor. The reference spur is measured to be at −75 dBc level indicating sufficient immunity of the PLL to the DSM switching activity. The PFD/CP linearization by operating it in a class-A mode also achieves better than 10 dB reduction of the nonlinearity-related spurious tones.

To sum up, spur cancellation is basically to cancel the analog transient distortion, and any efforts hit the fundamental limit of analog signal processing. Although

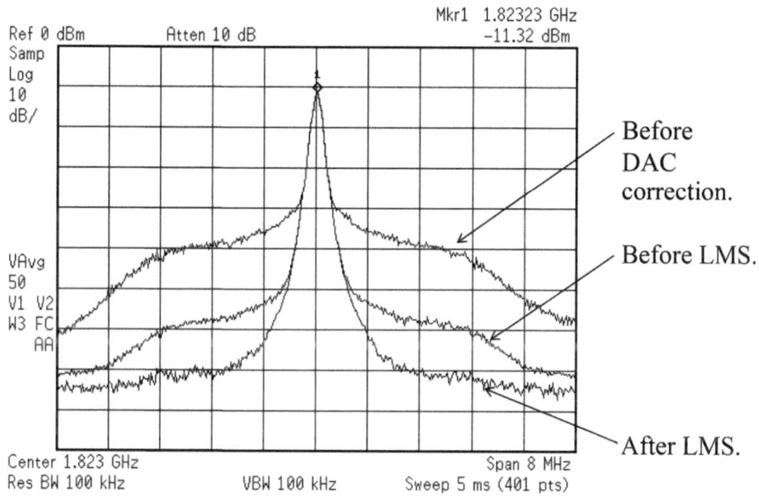

Fig. 6.38 RF spectrum before DAC correction and LMS

digital PLLs based on digital phase detection and loop filtering claim higher spur rejection, they are still limited by the distortion of the multi-bit DAC output. That is, it boils down to the same basic DAC issue of 1b vs. multi-bit DACs.

References

1. B. Song, CMOS RF circuits for data communications applications. IEEE J. Solid State Circuits **21**, 310–317 (1986)
2. F. Montaudon, R. Mina, S. Le Tual, L. Joet, D. Saias, R. Hossain, F. Sibille, C. Corre, V. Carrat, E. Chataigner, J. Lajoinie, S. Dedieu, F. Paillardet, E. Perea, A scalable 2.4-to-2.7GHz Wi-Fi/WiMAX discrete-time receiver in 65nm CMOS, *ISSCC Dig. Tech. Papers*, Feb. 2008, pp. 362–363
3. S. Lerstaveesin, M. Gupta, D. Kang, B. Song, A 48-860MHz CMOS low-IF direct conversion DTV tuner. IEEE J. Solid State Circuits **43**, 2013–2024 (2008)
4. T. Cho, D. Kang, C. Heng, B. Song, A 2.4GHz dual-mode 0.18μm CMOS transceiver for bluetooth and 802.11b. IEEE J. Solid State Circuits **39**, 1916–1926 (2004)
5. C. Lu, H. Wang, C. Peng, A. Goel, S. Son, P. Liang, A. Niknejad, H. Hwang, G. Chien, A 24.7dBm all-digital RF transmitter for multimode broadband applications in 40nm CMOS, *ISSCC Dig. Tech. Papers*, Feb. 2013, pp. 332–333
6. D. Chowdhury, S. Thyagarajan, L. Ye, E. Alon, A. Niknejad, Fully-integrated efficient CMOS inverse class-D power amplifier for digital polar transmitters. IEEE J. Solid State Circuits **47**, 1113–1122 (2012)
7. T. Sano, M. Mizokami, H. Matsui, K. Ueda, K. Shibata, K. Toyota, T. Saitou, H. Sato, K. Yahagi, Y. Hayashi, A 6.3mW BLE transceiver embedded RX image-rejection filter and TX harmonic-suppression filter reusing on-chip matching network, *ISSCC Dig. Tech. Papers*, Feb. 2015, pp. 240–241
8. M. Perrott, T. Tewksbury, C. Sodini, A 27-mW CMOS fractional-N synthesizer using digital compensation for 2.5-Mb/s GFSK modulation. IEEE J. Solid State Circuits **32**, 2048–2060 (1997)

9. W. Rhee, A. Ali, B. Song, A 1.1 GHz CMOS fractional-N frequency synthesizer with a 3-b 3rd-order delta-sigma modulator, *ISSCC Dig. Tech. Papers*, 2000, pp. 198–199
10. B. De Muer, M. Steyaert, On the analysis of delta-sigma fractional-*N* frequency synthesizers for high-spectral purity, *IEEE Trans. Circuits Syst. II*, Nov. 2003, pp. 784–793
11. S. Norsworthy, R. Schreier, G. Temes, *Delta-Sigma Data Converters: Theory, Design, and Simulation* (IEEE Press, New York, 1997)
12. B. De Muer, M. Steyaert, A CMOS monolithic delta-sigma-controlled fractional-*N* frequency synthesizer for DCS-1800. IEEE J. Solid State Circuits **37**, 835–844 (2002)
13. J. Craninckx, M. Steyaert, A fully integrated CMOS DCS-1800 frequency synthesizer. IEEE J. Solid State Circuits **33**, 2054–2065 (1998)
14. B. Miller, R. Conley, A multiple modulator fractional divider. IEEE Trans. Instrum. Meas. **40**, 578–583 (1991)
15. T. Riley, M. Copeland, T. Kwasniewski, Delta-sigma modulation in fractional-N frequency synthesis. IEEE J. Solid State Circuits **28**, 553–559 (1993)
16. G. Gillette, Digiphase synthesizer, in *Proceedings of 23rd Annual Frequency Control Symposium*, 1969, pp. 201–210
17. W. Rhee, A. Ali, An on-chip compensation technique in fractional-N frequency synthesis, in *IEEE International Symposium on Circuits and Systems*, 1999, pp. 363–366
18. I. Bietti, E. Temporitil, G. Albasini, R. Castello, An UMTS SD fractional synthesizer with 200kHz bandwidth and -128dBc/Hz @1MHz using spurs compensation and linearization techniques, in *IEEE Custom Integrated Circuits Conference*, 2003, pp. 463–466
19. S. Meninger, M. Perrott, A fractional-N frequency synthesizer architecture utilizing a mismatch compensated PFD/DAC structure for reduced quantization-induced phase noise, in *IEEE Transactions on Circuits and Systems II*, Nov. 2003, pp. 839–849
20. S. Pamarti, L. Jansson, I. Galton, A wideband 2.4GHz delta-sigma fractional-N PLL with 1-Mb/s in-loop modulation, IEEE J. Solid State Circuits **39**, 49–62 (2004)
21. Y. Dufour, Method and apparatus for performing fractional division charge compensation in a frequency synthesizer, U.S. patent 6 130 561 (2000)
22. Y. Koo et al., A fully integrated frequency synthesizer with charge-averaging charge pump and dual path loop filter fro PCS and cellular CDMA wireless systems, IEEE J. Solid State Circuits **37**, 536–542 (2002)
23. S. Pellerano et al., A dual band frequency synthesizer for 802.11a/b/g with fractional spur averaging technique, *ISSCC Dig. Tech. Papers*, 2005, pp. 20–22
24. M. Gupta, B. Song, A 1.8GHz spur-cancelled fractional-N frequency synthesizer with LMS-based DAC gain calibration, *ISSCC Dig. Tech. Papers*, Feb. 2006, pp. 478–479
25. M. Gupta, B. Song, A 1.8GHz spur-cancelled fractional-N frequency synthesizer with LMS-based DAC gain calibration. IEEE J. Solid State Circuits **41**, 2842–2851 (2006)
26. S. Dasgupta et al., Sign-sign LMS convergence with independent stochastic inputs. IEEE Trans. Inf Theory **36**, 197–201 (1990)

Chapter 7
Direct-Conversion Receivers

Old RF single-medium transceivers are mostly narrow-banded, and filtering at RF and IF helped to realize discrete RF systems. However, in modern integrated digital wireless systems, the lack of RF and IF filters heavily taxes on the performance of RF transceivers. In particular, channels with broadband spread spectrum and strong nearby blockers demand analog RF processing be extremely linear, and channel filtering highly selective. Direct frequency conversion to and from DC (Zero-IF) or low frequency (Low-IF) is the mathematical frequency translation using complex local (LO) carrier. On the transmitter (TX) side, key design factors are carrier leak, injection locking, out-of-band emission, spectral regrowth, and power amplifier (PA) efficiency. The carrier is suppressed using four-quadrant mixers, and the injection locking is avoided by dual conversion. The out-of-band emission is reduced below the spectral mask using RF filters or harmonic mixers, and the spectral regrowth issue doesn't exist with low IM3. The PA efficiency is also traded for linearity. On the receiver (RX) side, there are self-mixing, blocker, offset, harmonic mixing, and image issues. The self-mixing with the local carrier is avoided using low-IF or dual-conversion architectures, and the blocker is rejected by digital channel filters with wide dynamic range front-ends. The offset in the zero-IF receiver is eliminated using the feedback that is active only during the data packet period. The low-IF receiver exhibits no offset problem.

7.1 Direct or Dual Conversion

Standard super heterodyne RF receivers perform two frequency conversions using external RF image and IF channel filters. Although two VCOs are required, there exists no DC offset problem, and the requirements for IIP2 and IIP3 are relaxed due to the external high-Q filter. For integrated wireless systems, direct-conversion architectures are preferred since no external image and channel filters are required. They can be implemented with less number of RF components and one VCO, and

B.-S. Song, *System-level Techniques for Analog Performance Enhancement*,
DOI 10.1007/978-3-319-27921-3_7

low-pass filters can be used for channel selection. However, they suffer from DC offsets resulting from self-mixing and RF signal envelope, and significant baseband signal processing is required. The direct up-conversion is prone to the RF-to-LO injection locking and the carrier leak problems since the LO carrier is the same as the RF. It is difficult to generate a complex carrier since the VCO should oscillate at $2f_{RF}$ to perform divide-by-2.

On the other hand, dual-conversion architectures alleviate some of the problems such as reduced DC offset effect and no RF-to-LO injection locking problem in TX while sharing the same benefits of the direct-conversion system. However, the complexity grows with one more mixer stage. The first-stage LO1 is $(2/3)f_{RF}$, which is out of band, and the second-stage LO2 is obtained by dividing LO1 by two. That is, the complex carrier for LO2 can be easily obtained. Since the self-mixing gain is reduced by the LNA and mixer gain, the DC offset error gets smaller. Since local frequencies are lower, VCO phase noises are lower than the direct conversion. Image can be suppressed by using an on-chip LC filter in addition to the RF BPF. One of the benefits when using direct-conversion or dual conversion to zero-IF is that the image is its own self-image.

In direct-conversion receivers, IM2 generates the DC offset and beat frequencies of multiple interferers in the baseband. In dual-conversion receivers, the low-gain second-stage I/Q mixer allows the use of large overdrive voltage ($V_{gs} - V_{th}$), and reduces the IM2 mixing component. The IM2 component from the first RF-to-IF mixer moves out of the signal band. The IM2 of LNA leaks into the baseband after multiplied by the mixer offset. Differential LNA and mixers give high IP2, which is usually limited by the path mismatch of the differential signal. Two large adjacent channels create IM3 products in the desired channel band due to the nonlinearity of LNA, and IM3 components and thermal noise limit the SNR of the receiver. Therefore, careful biasing of the LNA input device is the most critical in RF designs. The $1/f$ noise corner frequency can be as high as 50 MHz at the LNA input, and 1 MHz at the baseband. The DC-offset cancellation at every burst of data packet suppresses the DC offset and slowly varying $1/f$ noise. Baseband filter/VGA opamps with PMOS input stages further reduce the $1/f$ noise effect on the baseband signal processing.

The choice of IF depends on the system bandwidth. Zero-IF is preferred for DC-free systems or wideband systems, but need static DC correction to avoid DC wandering. However, low-IF is better for narrowband systems. Direct conversion is ok for low-IF systems, but dual conversion is better for zero-IF systems.

7.2 Frequency Translation

The baseband signal is the low-pass spectrum centered at DC. To transmit or receive through the RF channel, the baseband spectrum should be translated in frequency to and from the band-pass spectrum at RF frequencies. The frequency up-conversion is carried out by modulating the amplitude, frequency, or phase

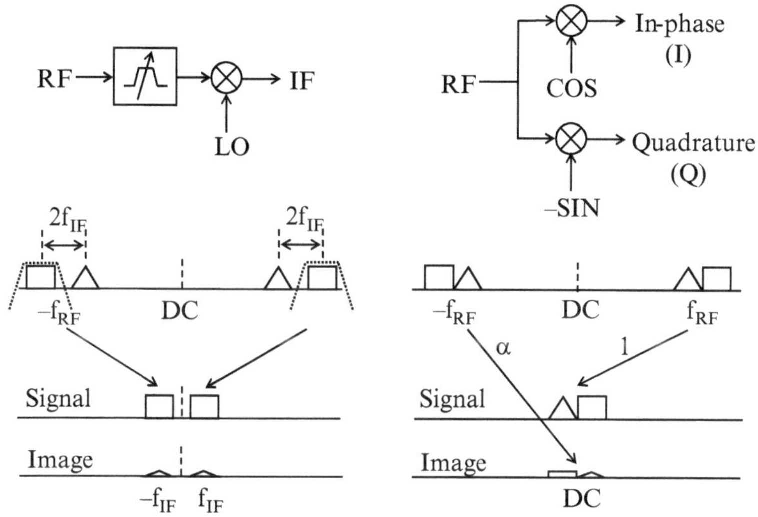

Fig. 7.1 Real and complex frequency down-conversions

(AM, FM, and PM) of the carrier frequency. The reverse down conversion requires an extraction of the carrier to recover the baseband signal for demodulation. Mixer performs the basic function of this frequency translation as shown in Fig. 7.1.

Mathematically, the mixer is a multiplier. Real mixer uses a real carrier of $\cos \omega_c t$ while complex mixer uses a complex carrier of $\exp(\omega_c t) = \cos \omega_c t + j\sin \omega_c t$. That is, the former translates frequency bi-directionally (shift right and left) by both $+\omega_c$ and $-\omega_c$ since $\cos \omega_c t = \{\exp(\omega_c t) + \exp(-\omega_c t)\}/2$. Due to this nature, the real mixer only allows the translation to and from the IF frequency, and two mixed frequencies can be separated by IF filtering. On the other hand, the complex in-phase (I) and quadrature (Q) mixers can translate frequency directly to DC and RF uni-directionally (shift right or left) by either $+\omega_c$ or $-\omega_c$. However, in reality, IF filters are not ideal, and bi-directional carriers are not matched. Therefore, the other directional frequency translation creates an image in the desired channel. Additionally, since mixer is made of a chopper, higher harmonics of the carrier yields messy harmonic mixing.

Following the SOC trend, wireless receivers evolve with more baseband digital processing at low- or zero-IF. Integrated RF receivers should meet a set of requirements quite different from those of the existing discrete receivers. Harmonic mixing in broadband systems such as TV tuners is critical as high-Q on-chip RF tracking filters are difficult to integrate, and image leakage in the low-IF down-conversion is inevitable. In the low-IF approach, typically 40–50 dB stronger adjacent or alternate channel becomes an image channel which no RF filter can suppress. Such low-IF receivers need an image rejection of higher than 55–65 dB to warrant 15 dB SNR. On the other hand, in the zero-IF approach, the image problem is not as serious since the image is its own self-image.

Although using zero-IF greatly simplifies the receiver architecture for networking or OFDM systems, the well-known DC and 1/*f* noise problem should be solved. Because the sensitivity test condition is specified with a constant input only, even zero-IF systems can meet the specification. However, the DC instability makes it very difficult to detect data at the sensitivity point when the power envelope of the received signal varies rapidly with strong AM modulation. DC wanders due to the high-pass filters added in the *I*/*Q* branches to block the DC offset and 1/*f* noise, and any sudden DC transients die out with long time constants of the AC-coupling networks. The DC wandering problem can be avoided by adopting the low-IF architecture.

7.2.1 Harmonic Mixing and Image Folding

The benefit of using a passive RF tracking filter is so obvious. It pre-filters blocker, image, and harmonic mixing components before any mixing. Although low-*Q* RF filters cannot remove all blockers, the mixer linearity and dynamic range requirements are greatly relieved since far-away blockers are filtered to some degree. Integrated RF systems rely on the complex down-conversion with *I*/*Q* mixers. Unlike the real signal, which has both positive and negative frequencies, the complex signal has only one frequency, either positive, $\exp(\omega t)$, or negative, $\exp(-\omega t)$. Therefore, if a complex LO is used to down-convert a complex RF signal, the harmonic mixing and image folding work differently from the real mixing case. In the complex *I*/*Q* mixing, the complex local carrier is ideally a single tone at $-f_{\mathrm{RF}}$.

In the chopper mixer, due to the inherent current switching controlled by the local carrier, the mixing operation is equivalent to the multiplication by a square-wave local carrier as follows, which down-converts odd harmonic bands into the baseband.

$$V_{\mathrm{LO}}(t) = \frac{4}{\pi} \times \left(\sin \omega_{\mathrm{LO}} t + \frac{1}{3} \times \sin 3\omega_{\mathrm{LO}} t + \frac{1}{5} \times \sin 5\omega_{\mathrm{LO}} t + \frac{1}{7} \times \sin 7\omega_{\mathrm{LO}} t + \cdots \right). \tag{7.1}$$

The complex LO carrier has the odd harmonics with magnitudes of 1/3, 1/5, 1/7, ... like the real square-wave carrier, but the harmonics show up alternately at positive and negative frequencies like $+3f_{\mathrm{LO}}$, $-5f_{\mathrm{LO}}$, $+7f_{\mathrm{LO}}$, and so on as illustrated in Fig. 7.2.

The ideal complex *I*/*Q* down-conversion is the case when both the complex input and the complex LO carrier have real and imaginary parts as shown in Fig. 7.3.

Four real multipliers are required to get the four real product terms. In the ideal complex down-conversion, only positive or negative odd harmonic channels are mixed and folded down into the baseband as shown in Figs. 7.4 and 7.5.

All odd LO harmonics of the 3rd, 7th, 11th, ... do not pick up negative-frequency channels as they do not exist. The channel at $+3f_{\mathrm{LO}}$ is moved to $+2f_{\mathrm{LO}}$,

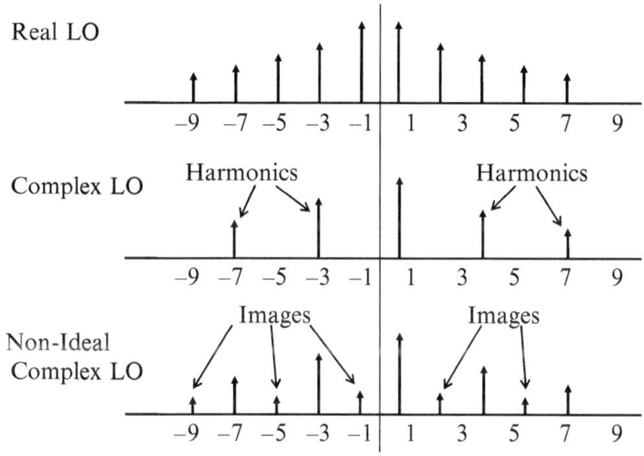

Fig. 7.2 Harmonic and image components of LO carriers

Fig. 7.3 Ideal complex down-conversion

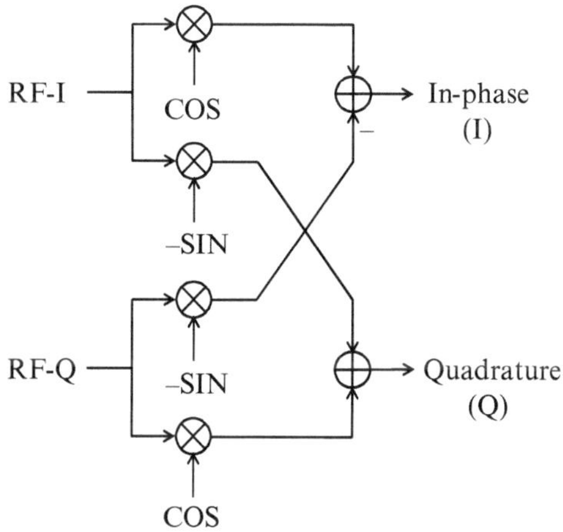

which is out of the low-IF band. Furthermore, there is no image folded into the low-IF band because the complex operation of multiplying by the complex $\exp(-\omega_{\mathrm{LO}}t)$ is just to translate the frequency by $-f_{\mathrm{LO}}$. However, in practice, both harmonic mixing and image problems arise when the complex multiplier which emulates the four real multipliers are not ideal as shown in Fig. 7.6.

The negative-frequency LO carrier and its harmonics have positive-frequency leaks attenuated by α. Similarly, the positive-frequency RF and its harmonics have negative-frequency leaks attenuated by β. That is, the local I/Q mismatch creates the image of the LO and its harmonic components proportional to a mismatch factor α at $+f_{\mathrm{LO}}$, $-3f_{\mathrm{LO}}$, $+5f_{\mathrm{LO}}$, and $-7f_{\mathrm{LO}}$, etc... In addition, the non-ideal complex RF signal has a negative-frequency spectrum leaking due to the I/Q path mismatch factor β. Therefore, in the non-ideal complex I/Q down-conversion, numerous undesired harmonic and image bands are mixed down into the desired signal

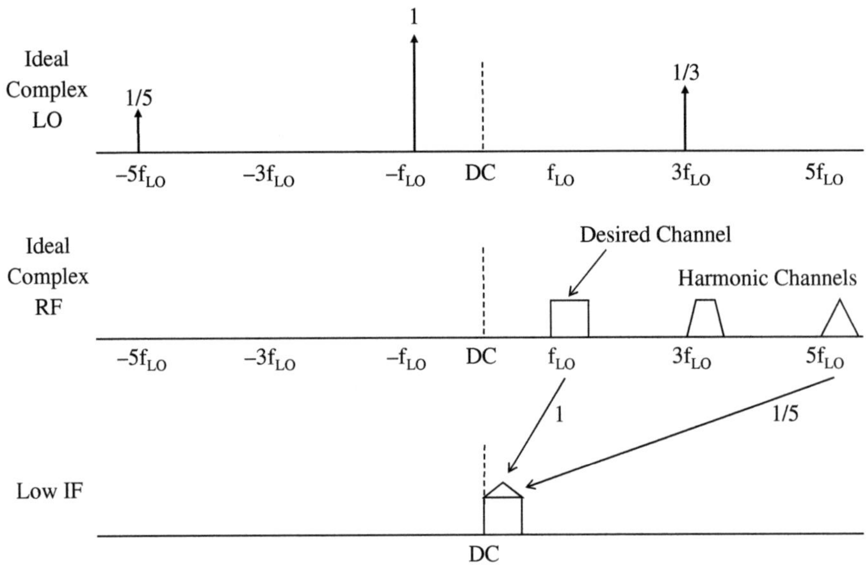

Fig. 7.4 Harmonic mixing in ideal complex down-conversion

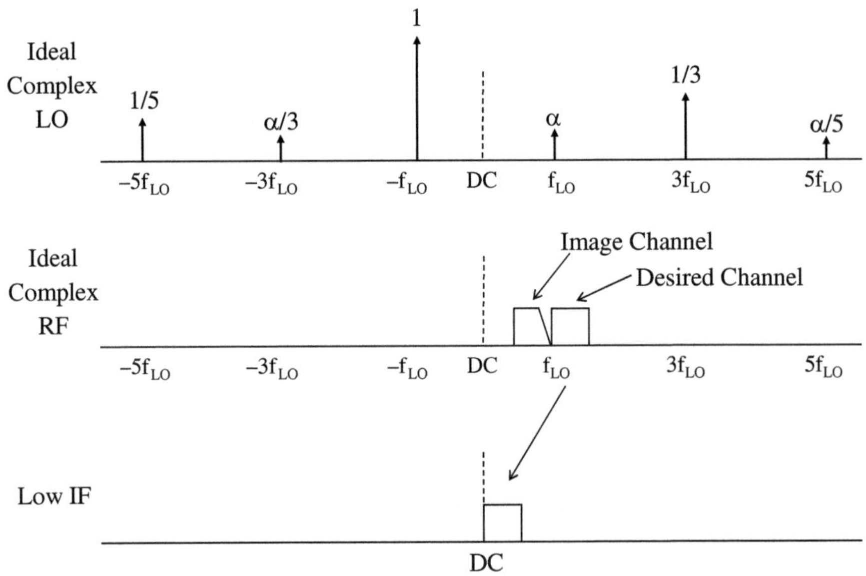

Fig. 7.5 No image folding in ideal complex down-conversion

band. For example, the third-harmonic channel and its image alias into the low-IF band after attenuated by $\alpha/3$ and $\beta/3$, respectively. That is, the complex down-conversion is not completely free of the 3rd-harmonic mixing, and it is necessary to filter the 3rd-harmonic band before down-conversion.

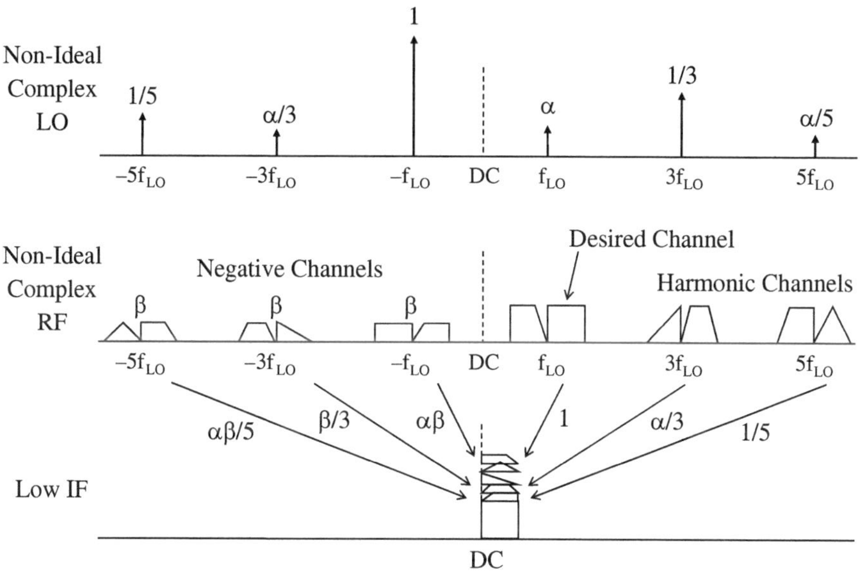

Fig. 7.6 Non-ideal complex down-conversion

The complex mixing has been shown to be superior in rejecting harmonic and image components though limited by the non-idealities of the four real mixers. In digital wireless transmitters, the complex baseband *I*/*Q* signals are often generated digitally. However, in most RF receivers, the RF signal is real, and has no imaginary component unless RF poly-phase or Hilbert filters are used. That is, in the down-conversion case, the RF input is only real though the LO carrier can be complex. Such real input case works similarly as in the complex down-conversion except for the fact that the negative-frequency RF channel which is as strong as the desired positive-frequency channel. Both harmonic mixing and image folding situations are explained in Fig. 7.7.

For real RF signals, the positive and negative spectral densities are the same, and the dominant harmonic mixing is from the third-harmonic channel at $-3f_{RF}$ while the image is from the RF at $-f_{RF}$. Their magnitudes relative to the down-converted signal are defined as harmonic and image rejections, respectively.

7.2.2 Harmonic Rejection

In traditional high-IF super heterodyne receivers, RF tracking filter suppresses harmonic components even before down-conversion mixing. However, in zero- or low-IF receiver, many approaches are taken to avoid harmonic mixing, but no clear solutions stand out. The up/down double conversion is to mix the RF spectrum up to much higher frequencies and to mix it down after filtering coarsely. If double

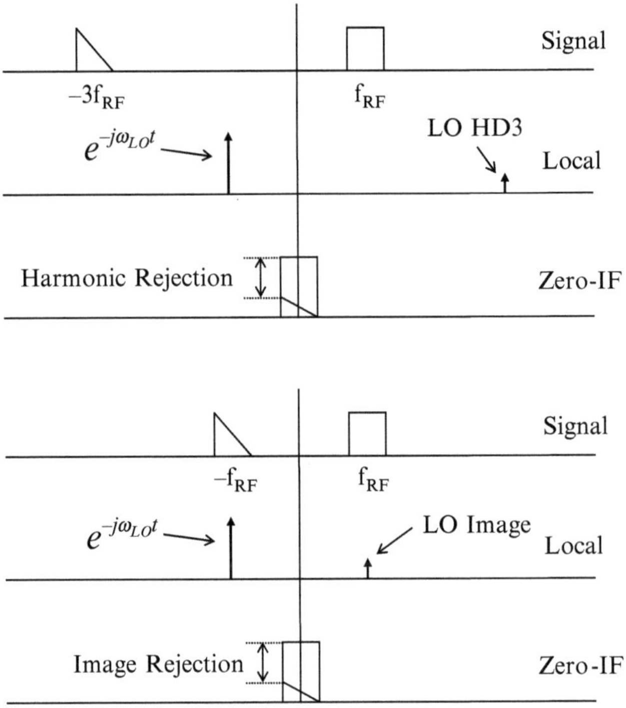

Fig. 7.7 Down-conversion of real RF

quadrature (complex) mixer is used, the third harmonic is rejected. The most common way is to use both a coarse RF tracking filter and a poly-phase harmonic rejection mixer.

Due to the lack of high-Q tuning elements in CMOS, the selectivity achievable with on-chip tunable RF band-pass filters is limited, and their power consumption will be prohibitively high. Taking this for granted, the mixing effect by odd LO harmonics can be further reduced using poly-phase mixers. Poly-phase mixing emulates the sinusoidal mixing by using the sum of multiple poly-phase LO vectors to cancel the effect of the harmonics. The principle of poly-phase mixing is that the signal and its harmonics change their phase at different rates. For example, the third harmonic changes its phase three times faster than the fundamental. This implies that the poly-phase mixer outputs are added constructively and destructively. The phase of the 3rd harmonic is shown to change three times faster than that of the fundamental in Fig. 7.8.

If the three phases of the fundamental LO vectors are 0°, 45°, and 90°, those of the 3rd-harmonic LO vectors are 0°, 135°, and 270°, respectively. Therefore, the three-phase 3rd-harmonic LO vectors are destructively summed to be zero. So are the three-phase 5th-harmonic LO vectors. This implies that the poly-phase mixing can prevent channels at certain LO harmonics from being down-converted, but the

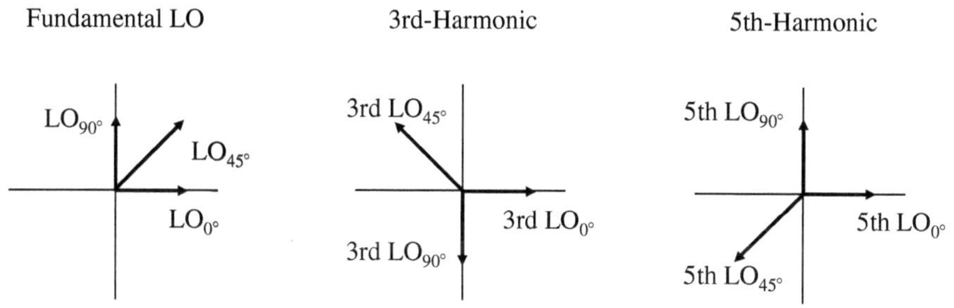

Fig. 7.8 Sums of three vectors

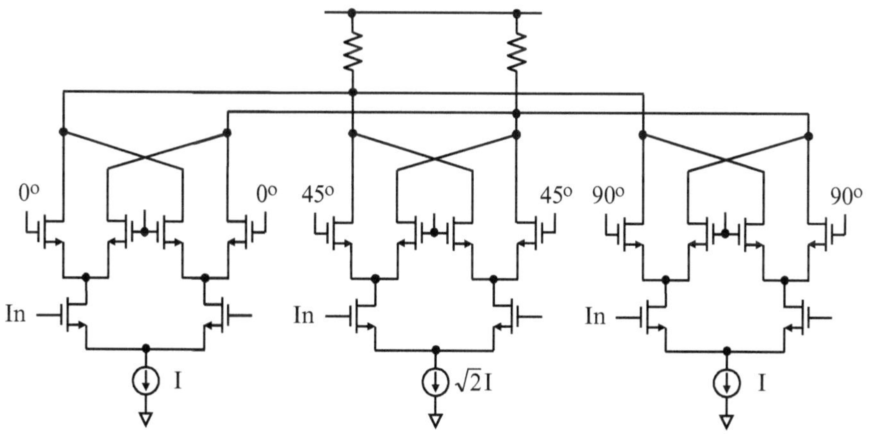

Fig. 7.9 Three-phase poly-phase mixer

gain variations and phase errors in the multiple mixer paths limit the upper bound of the achievable suppression of the harmonic mixing components. That is, the ideal complex *I*/*Q* poly-phase mixing also eliminates the image leakage, but the image suppression is similarly limited by the same *I*/*Q* path mismatch.

The three-phase mixing example is shown in Fig. 7.9. The complex *I*/*Q* three-phase harmonic-reject mixer is implemented by splitting a chopper-type mixer into three individual mixer cells with a gain ratio. The ideal three-phase mixer can eliminate the harmonics of the LO at 3rd, 5th, 11th, 13th, and so on. It is to replicate the sinewave DAC with three taps made of three chopper mixers. Since the sum of the output currents is equivalent to the product of the RF and sinusoidal LO carriers, the mixer noise and nonlinearity are affected by the dynamic transient non-ideality of the switching chopper mixer as in the DAC. If the gain ratio is truncated approximately to be (1, 1.41, 1), the 3rd and 5th-harmonic rejections are about 66 and 71 dB, respectively. In theory, poly-phase mixing with many chopper mixer taps is an asymptote to the sinusoidal mixing, but the performance is limited due to other practical reasons. For example, the suppression of the 3rd and 5th-harmonics ends up with about 40 dB if the gain and phase errors in the *I*/*Q* branches from the RF filter, three-phase LO, and three mixer cells are on the order of 1 % and 0.5°, respectively.

7.2.3 Image Rejection

Unlike the harmonic mixing problem with far-away channels, the image problem with nearby channels is rather cumbersome. In the zero-IF receiver, the image is of its own, and manageable. However, in the low-IF receiver, any strong adjacent, alternate, or nearby blockers are folded into the baseband. Many analog solutions such as double quadrature mixer, complex $\Delta\Sigma$ modulator, complex poly-phase filter, and *I*/*Q* oscillator have been tried [1–4]. Although they suppress images generated by specific local circuit elements, analog methods still rely on *I*/*Q* path gain and phase matching, and are not effective since image at the system level is not cancelled at all.

Analog calibration schemes have been reported to get higher image rejection than achievable with simple analog *I*/*Q* matching methods. A variable gain amplifier and a phase shifter are used to correct *I*/*Q* path gain and phase errors, but it is difficult to track drifting parameters constantly using offline calibration [5]. Time-varying errors can only be canceled effectively using online calibration or compensation methods that operate all the time [6]. Gain and phase errors are estimated by evaluating both signal and image responses of a test tone, and *I*/*Q* paths are compensated digitally at low IF [7]. The test tone is also used to correct the phase error, which is detected by integrating the product of *I*/*Q* signals [8]. Image is rejected using an externally controlled AGC for gain correction and a BER optimizer for phase correction [9]. Among all image rejection methods reported to date, the most general one is the post digital processing method for *I*/*Q* imbalance correction [10–12], which separates signal from image using symmetric adaptive de-correlation [13] and least-mean-square (LMS) algorithm [14].

The symmetric signal de-correlation method is based on the simple fact that signal and image are uncorrelated. Due to the symmetry in the circuit, image leaks into the signal band, and similarly signal leaks into the image band. This symmetric leakage can be modeled using two small complex coefficients. Frequency-dependent multi-tap LMS adaptation for these two coefficients is conceptually very general, but it suffers from phantom solutions [13]. In particular, the algorithm fails to converge if the signal is made of strong discrete tones such as TV video and sound carriers. Multiple frequency-dependent taps can be updated correctly only if the signal occupies the full bandwidth of interest [10]. On the other hand, single-tap adaptation based on an assumption that the two leakage coefficients are complex conjugates of each other performs as two identical LMS adaptation loops, and converges well for all signal conditions [11]. As a variation, asymmetric one-loop adaptation and also sharing computational resources can save power and area [12].

The simplest I/Q error correction method is based on a zero-forcing adaptive feedback concept using a sign–sign LMS error update algorithm [15, 16]. The complex image rejection algorithm is reformulated using real signal processing blocks. The gain and phase errors are detected by using the orthonormal nature of I/Q signals, and the error updating algorithm is simplified to rely on four sign detections. The two key functional blocks that need to be added for image rejection are only an image rejecter and an error detector, which can be implemented either in the analog or the digital domain.

As is true for all digital methods, image rejection is ultimately limited by the quantization noise of I/Q ADCs if the image rejecter is implemented in the digital domain. Most RF systems have high-resolution ADCs with wide dynamic range to quantize signal faithfully in the presence of large blockers. Therefore, digital-domain image rejection down to the level of quantization error would suffice in most systems. The accuracy of analog image rejection is restricted by the DC offsets of the analog comparators used to detect the sign and I/Q path mismatch. One possible variation is a hybrid form that implements the image rejecter in the analog domain but the error detector digitally. For example, an analog complex baseband S/H can be used as an analog image rejecter. The image rejection ratio (IRR) of such a hybrid system is not limited by the resolution of the ADCs but by the trimming accuracy of the analog image rejecter. When applied to any direct-conversion low- or zero-IF receivers, the baseband image rejection algorithm can suppress the total accumulated image at the system level irrespective of its sources with simple post baseband processing. Unlike most schemes that require elaborate complex I/Q analog circuits, the improvement in image rejection results entirely from the zero-forcing adaptive LMS feedback. Therefore, such image rejection is suitable for system integration, and can greatly simplify overall RF receiver architectures.

7.3 Image in Direct-Conversion Receivers

In super heterodyne receivers, incompletely filtered components in the image band leak into the IF band. Passive filters such as LC or surface acoustic wave (SAW) filters that have a high-Q have been commonly used to filter the RF signal. Due to the difficulty of achieving such high selectivity, most existing RF systems set the IF usually high at about 1/10 to 1/20 of the RF (10.7 MHz for FM radio and 36/44 MHz for TV). However, the direct down-conversion to zero- or low-IF does not allow external high-Q RF filters. It is a frequency translation operation, which shifts the whole received spectrum down without impairments using a complex negative-frequency local carrier. This mixing process represented by a complex local carrier is equivalent to multiplying the real RF signal by two real I/Q carriers with a 90° phase difference.

7.3.1 Complex Image

Two down-converted I/Q signals at low or zero-IF represent the real and imaginary parts of the complex baseband signal. The complex baseband signal $I+jQ$ is the positive-frequency signal moved down to the IF while the complex image $I-jQ$ is the negative-frequency image moved up to the IF. Note that $I-jQ$ is the self-image of $I+jQ$ in the zero-IF case, but two complex conjugate vectors are still uncorrelated since they rotate in the opposite directions [11]. Only the desired complex signal $I+jQ$ is down-converted if these I/Q signals are perfectly matched in magnitude and 90° out-of-phase. In practice, imbalances in the I/Q path gain and phase cause the complex image to leak into the desired down-converted baseband. This image leakage can be better understood if the complex local carrier is assumed to have a positive-frequency component that is α times smaller than the desired negative-frequency carrier. The magnitude of this small complex leakage coefficient α is defined as IRR. The non-ideal complex carrier is now represented by

$$V_{\mathrm{LO}}(t) = e^{-i\omega_{\mathrm{LO}}t} + \alpha \times e^{-i\omega_{\mathrm{LO}}t}, \tag{7.2}$$

implying that the small component of the positive-frequency carrier picks up the complex image. All images in the down-conversion process are shown in Fig. 7.10.

Consider a complex channel with a path gain mismatch of a and a phase mismatch of θ. Let both a and θ be constants independent of frequency. This assumption is valid since I/Q mixers, AGCs, anti-aliasing filters, and ADCs found in most digital RF receivers have wide bandwidths, and digital I/Q channel filters can be designed to be exact. However, a and θ can be frequency-dependent in systems with analog channel filters. The image rejection algorithm loses some

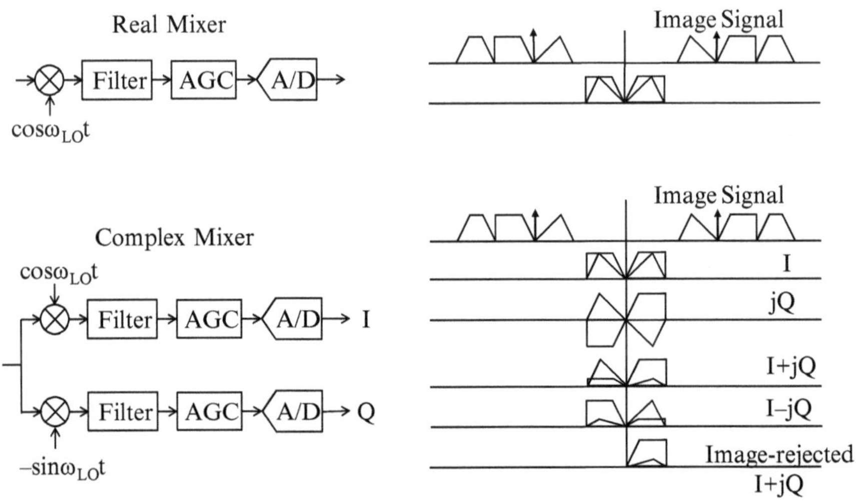

Fig. 7.10 Images in down-conversion

accuracy in such cases. In a single-tone example, let's redefine the non-ideal *I*/*Q* signals as I' and Q'.

$$I' = \left(1+\frac{a}{2}\right)\cos\left(\omega t+\frac{\theta}{2}\right) = \left(1+\frac{a}{2}\right)\left\{\frac{e^{j\left(\omega t+\frac{\theta}{2}\right)}+e^{-j\left(\omega t+\frac{\theta}{2}\right)}}{2}\right\}. \tag{7.3}$$

$$Q' = \left(1-\frac{a}{2}\right)\sin\left(\omega t-\frac{\theta}{2}\right) = \left(1-\frac{a}{2}\right)\left\{\frac{e^{j\left(\omega t-\frac{\theta}{2}\right)}-e^{-j\left(\omega t-\frac{\theta}{2}\right)}}{2j}\right\}. \tag{7.4}$$

Then, assuming a and θ are small, the non-ideal complex signal $I'+jQ'$ and the non-ideal complex image $I'-jQ'$ can be approximated using Taylor series.

$$I'+jQ' \approx e^{j\omega t}+\left(\frac{a-j\theta}{2}\right)e^{-j\omega t}. \tag{7.5}$$

$$I'-jQ' \approx \left(\frac{a+j\theta}{2}\right)e^{j\omega t}+e^{-j\omega t}. \tag{7.6}$$

It can be understood that there exist symmetric leakages between the complex signal and the complex image. Note that the definitions of signal and image are interchangeable. That is, $I'+jQ'$ is approximately the ideal $I+jQ$ plus $I-jQ$ multiplied by $(a-j\theta)/2$. Similarly, $I'-jQ'$ is approximately the ideal $I-jQ$ plus $I+jQ$ multiplied by $(a+j\theta)/2$. If the image leakage coefficient α is defined as $(a-j\theta)/2$, then the signal leakage coefficient α^* is its complex conjugate $(a+j\theta)/2$.

That is, if the image leakage coefficient α is known, a complex image rejecter can be implemented as shown on the left side of Fig. 7.11. Therefore, an image rejecter for real *I*/*Q* channels can be derived as shown on the right side.

7.3.2 Complex Image Rejection Algorithm

I/*Q* channels with gain and phase errors can be modeled as a non-ideal complex channel in a matrix form. I' and Q' defined by (7.3) and (7.4) are shown in Fig. 7.12.

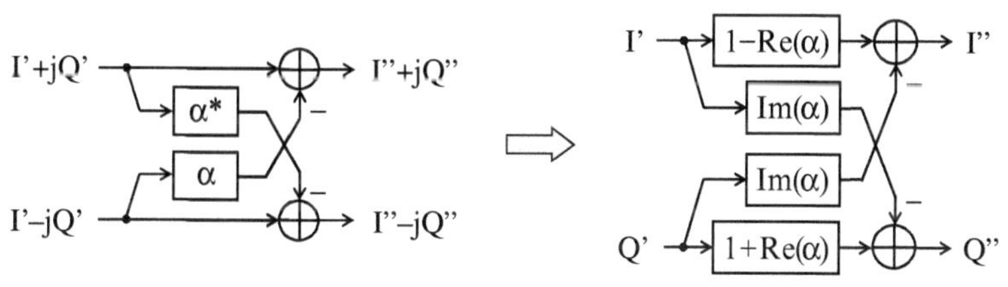

Fig. 7.11 Complex image rejection concept

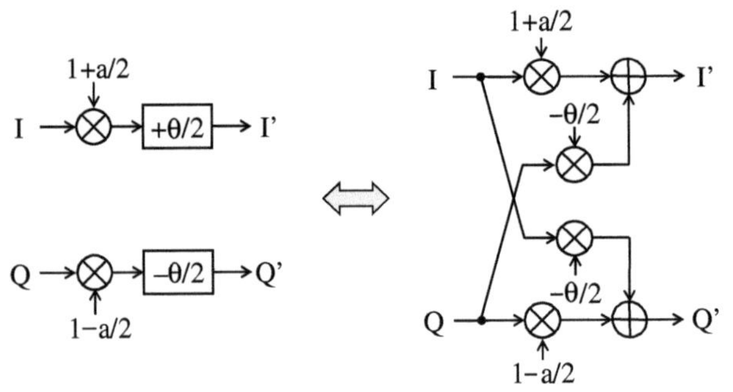

Fig. 7.12 Non-ideal complex channel and its linear model

Since the errors are small, the small-signal non-ideal channel transfer function can be derived as follows.

$$\begin{bmatrix} I' \\ Q' \end{bmatrix} = \begin{bmatrix} 1+\frac{a}{2} & -\frac{\theta}{2} \\ -\frac{\theta}{2} & 1-\frac{a}{2} \end{bmatrix} \times \begin{bmatrix} I \\ Q \end{bmatrix}. \tag{7.7}$$

It implies that I' and Q' are approximated by a linear combination of I and Q as shown on the right side. The process of rejecting the image is to recover the ideal orthonormal I/Q signals from I' and Q'. That is, by undoing what the non-ideal complex channel did to I/Q signals, the original complex signal can be restored. This is possible by inverting the matrix of the channel transfer function. Therefore, the image rejecter performs the following function.

$$\begin{bmatrix} I'' \\ Q'' \end{bmatrix} = \begin{bmatrix} 1-\frac{a}{2} & \frac{\theta}{2} \\ \frac{\theta}{2} & 1+\frac{a}{2} \end{bmatrix} \times \begin{bmatrix} I' \\ Q' \end{bmatrix}. \tag{7.8}$$

This concept is graphically explained in Fig. 7.13.

The image-free outputs from the image rejecter are I'' and Q'', but they contain higher-order terms of a and θ. If the third- and higher-order terms are neglected, I'' and Q'' are orthonormal and very close to the ideal I and Q as approximated below.

$$\begin{bmatrix} I'' \\ Q'' \end{bmatrix} = \begin{bmatrix} 1-\frac{a^2}{4}-\frac{\theta^2}{4} & \frac{-a\theta}{4} \\ \frac{a\theta}{4} & 1-\frac{a^2}{4}-\frac{\theta^2}{4} \end{bmatrix} \times \begin{bmatrix} I \\ Q \end{bmatrix} \approx \begin{bmatrix} 1 & 0 \\ 0 & 1 \end{bmatrix} \times \begin{bmatrix} I \\ Q \end{bmatrix} = \begin{bmatrix} I \\ Q \end{bmatrix}. \tag{7.9}$$

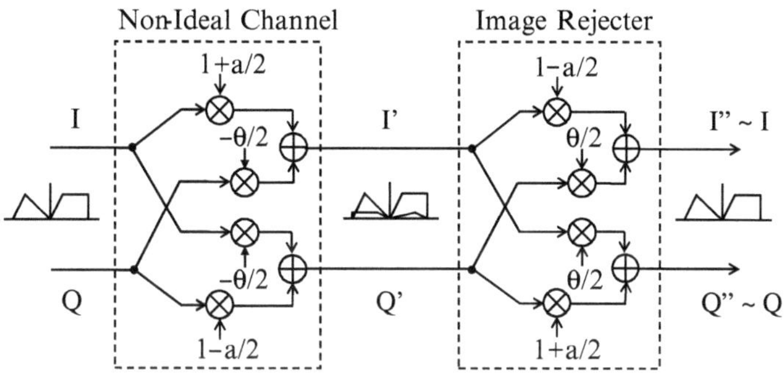

Fig. 7.13 Image model and rejection algorithm

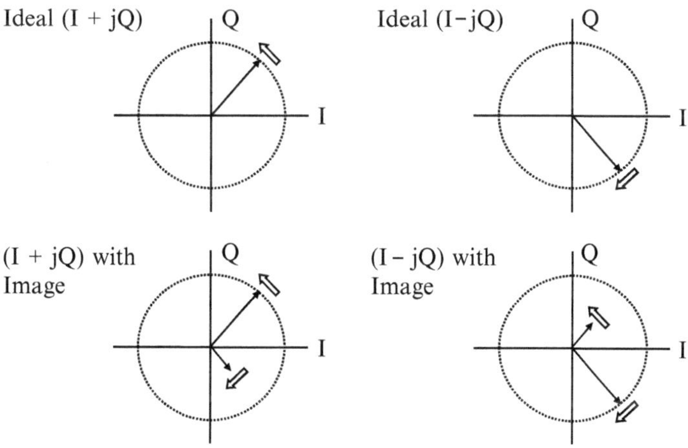

Fig. 7.14 Two vectors rotating in opposite directions

The values of a and θ should be known accurately for the correct operation of this image rejection scheme. Two vectors of $I+jQ$ and $I-jQ$ are rotating in opposite directions as sketched with and without image in Fig. 7.14. Since the image is the leaked opposite-directional vector, the resultant vector has both magnitude and phase errors. They are uncorrelated, and can be advantageously used to correlate out the image component in the signal or the signal component in the image.

The I/Q vectors before and after image rejection are also conceptually sketched in Fig. 7.15. The resulting orthonormal vectors are perpendicular to each other with equal magnitudes. So there is no image of one vector left on the other vector.

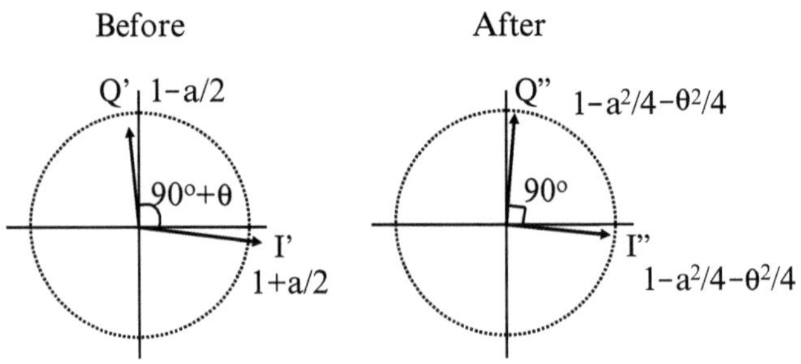

Fig. 7.15 *I/Q* vectors before and after image rejection

7.3.3 Path Gain and Phase Error Detector

Small residual gain and phase errors still exist in the image-rejected I'' and Q'' signals unless their exact values are known in the inverted channel response. These residual errors can be reduced adaptively by applying a zero-forcing feedback loop. The gain error a can be estimated by low-pass filtering $I^2 + Q^2$ as follows.

$$\mathrm{LPF}\left(I^2 - Q^2\right) = \frac{\left(1 + \frac{a}{2}\right)^2}{2} - \frac{\left(1 - \frac{a}{2}\right)^2}{2} \approx a. \tag{7.10}$$

The low-pass filtered $I^2 + Q^2$ converges to zero if I and Q have the same magnitude. Similarly, the phase error can be obtained using the orthogonal property of I/Q signals. Low-pass filtering the product of I and Q yields the phase error of $-\theta/2$.

$$\mathrm{LPF}(IQ) = -\frac{\theta}{2} \times \left(1 - \frac{a^2}{4}\right) \approx -\frac{\theta}{2}. \tag{7.11}$$

The low-pass filtered I/Q product will also approach zero if I and Q are exactly 90° out-of-phase. The above definitions of gain and phase errors are rather intuitive. The same conclusion can be drawn even by applying the signal de-correlation principle directly. To de-correlate the signal from the image, the image component in $I + jQ$ is correlated with $I - jQ$. That is, $(I + jQ)(I - jQ)^* = I^2 - Q^2 + 2jIQ$. The residual complex error should be averaged to be $a - j\theta$, which gives the same results as derived. This conclusion, drawn for the single-tone case, holds true for any amplitude-, frequency-, or phase-modulated signals. To avoid the complexity of estimating these gain and phase errors accurately, only the signs of the errors can be detected for LMS adaptation.

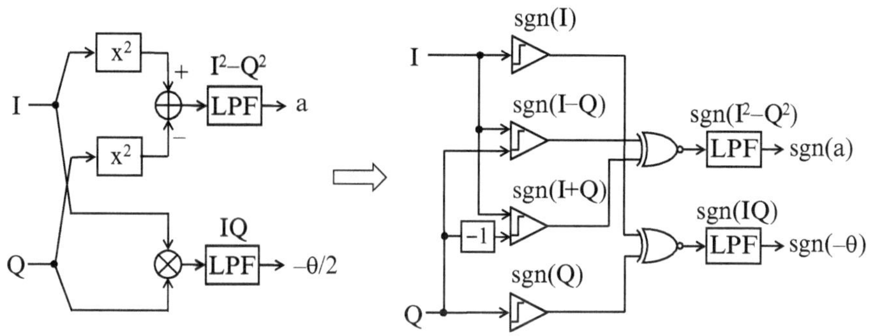

Fig. 7.16 Sign–sign path gain and phase error detection

7.3.4 *Sign–Sign LMS Algorithm for Image Rejection*

In the complex symmetric signal de-correlation technique, eight multiplications and eight additions are needed to estimate gain and phase errors although the asymmetric adaptation can be simplified to use three multiplications and three additions. In the sign-detection approach, however, only four sign detectors and two XNOR gates are needed as shown in Fig. 7.16.

Note that the sign of $I^2 - Q^2$ can be obtained by XNORing the signs of $I+Q$ and $I-Q$, and similarly, the sign of IQ by the signs of I and Q. The errors can be updated using the following two standard LMS-based sign–sign algorithms.

$$a[k+1] = a[k] + \mu_a \times \mathrm{sgn}[\mathrm{LPF}\{\mathrm{sgn}(I+Q) \times \mathrm{sgn}(I-Q)\}]. \tag{7.12}$$

$$\theta[k+1] = \theta[k] - \mu_\theta \times \mathrm{sgn}[\mathrm{LPF}\{\mathrm{sgn}(I) \times \mathrm{sgn}(Q)\}]. \tag{7.13}$$

Thus, if the low-pass filtered sgn($I^2 - Q^2$) is positive, a is increased by μ_a. Similarly, if the low-pass filtered sgn(IQ) is positive, the parameter θ is decreased by μ_θ. These step sizes should be made small enough to achieve high IRR, but too small steps slow down the adaptation process.

The direction of the gain and phase error update is plotted in Fig. 7.17. It is shown that the solutions of a and θ converge to the correct values regardless of the initial values in the two-dimensional plane. Let $e_a[k]$ and $e_\theta[k]$ be the errors detected for a and θ, which are defined as the difference between the ideal value and the current estimate, for further analysis on stability and convergence. The adaptation equations can be written using $e_a[k]$ and $e_\theta[k]$ as follows.

$$a[k+1] = a[k] + \mu_a \mathrm{sgn}(e_a[k]). \tag{7.14}$$

$$\theta[k+1] = \theta[k] - \mu_\theta \mathrm{sgn}(-e_\theta[k]). \tag{7.15}$$

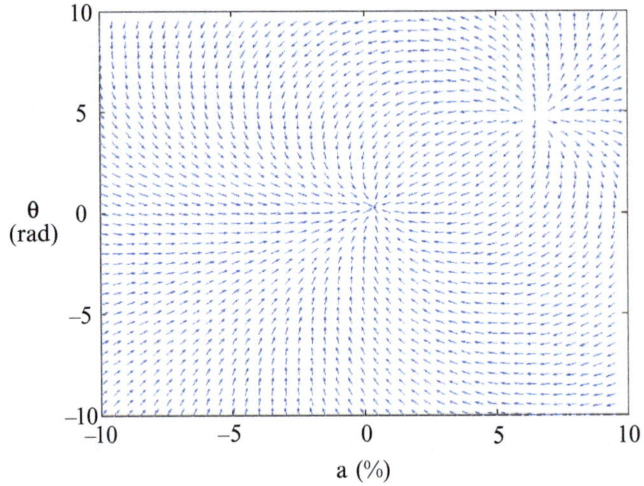

Fig. 7.17 Direction of gain and phase error update

Fig. 7.18 Recursive function of the detection error

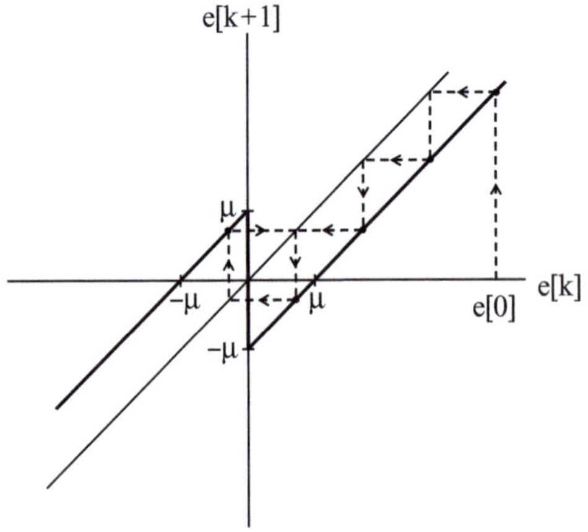

The newly detected errors, $e_a[k+1]$ and $e_\theta[k+1]$, can be rewritten in a self-recursive form.

$$e_a[k+1] = a - a[k+1] = e_a[k] - \mu_a \mathrm{sgn}(e_a[k]). \tag{7.16}$$

$$e_\theta[k+1] = \theta - \theta[k+1] = e_\theta[k] - \mu_\theta \mathrm{sgn}(e_\theta[k]). \tag{7.17}$$

An example of the recursive function is shown in Fig. 7.18. The detected error follows the dashed trajectory. The error gets smaller in the next update if the magnitude of the detected error is greater than μ. Once the detected error $e[k]$ falls within the error bound of $\pm\mu$, then $e[k+2] = e[k]$. Thus the estimated gain and

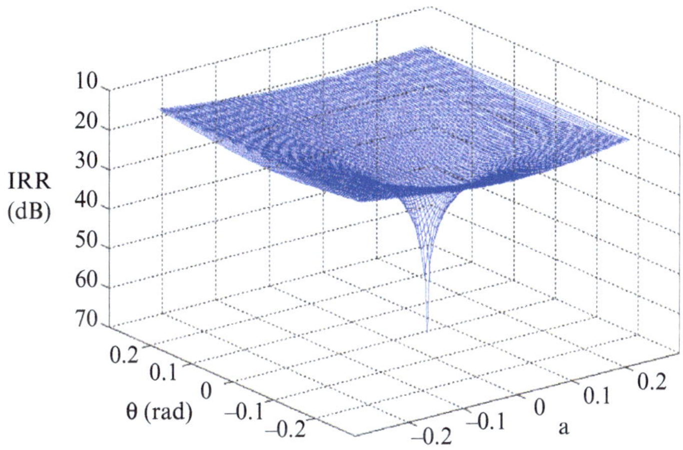

Fig. 7.19 IRR vs. gain and phase errors

phase errors will converge to the correct values within $\pm\mu$ irrespective of the initial condition.

The IRR is plotted as a function of a and θ in Fig. 7.19. This three-dimensional plot shows that the IRR is maximized if a and θ converge to the correct values. There is only one minimum error point, and there are no other local minima.

7.3.5 *Magnitude vs. Sign Detection*

The sign–sign error detection algorithm is based only on the polarities of $I, Q, I+Q$ and $I-Q$ instead of their magnitudes, and allows simple hardware. However, the sign detection only may be inferior to the magnitude detection in terms of error estimation accuracy. As an example, LPF{sgn($I-Q$)*sgn($I+Q$)} should be LPF {(I^2-Q^2)} if actual I and Q magnitudes are considered. Assuming that I^2-Q^2 is Gaussian distributed with a mean of a and a variance of σ^2, averaging N samples leads to

$$\frac{1}{N}\sum_{i=1}^{N}(S+a), \tag{7.18}$$

where $S=N(0,\sigma^2)$. This is a random variable with a mean of a and a variance of σ^2/N. The update direction is correct with a probability of

$$P = 1 - Q\left(\frac{|a|\sqrt{N}}{\sigma}\right), \quad \text{where } Q(X) = \frac{1}{\sqrt{2\pi}}\int_X^\infty e^{-\frac{x^2}{2}}dx. \tag{7.19}$$

With the sign detection, it becomes

$$\frac{1}{N}\sum_{i=1}^{N}\mathrm{sgn}(S+a). \tag{7.20}$$

Each quantity in the sum has a mean value of $1-2Q(a/\sigma)$. Then its variance is defined as

$$\begin{aligned}\mathrm{Var}\left\{\frac{1}{N}\sum_{i=1}^{N}\mathrm{sgn}(S+a)\right\} &= \frac{1}{N}\left(E\{\mathrm{sgn}(S+a)\}^2-[E\{\mathrm{sgn}(S+a)\}]^2\right)\\ &= \frac{1}{N}\left[1-\left\{1-2Q\left(\frac{a}{\sigma}\right)\right\}^2\right]=\frac{1}{N}\left\{4Q\left(\frac{a}{\sigma}\right)-4Q^2\left(\frac{a}{\sigma}\right)\right\}.\end{aligned} \tag{7.21}$$

If the gain error a is assumed to be small, $Q(a/\sigma)$ can be approximated as

$$Q(a/\sigma)=0.5-\frac{a}{\sqrt{2\pi}\sigma}. \tag{7.22}$$

The mean and the variance are approximately

$$\sqrt{\frac{2}{\pi}}\frac{a}{\sigma}\quad\text{and}\quad\frac{1}{N}, \tag{7.23}$$

respectively. Similarly, the update direction is correct with a probability of

$$P=1-Q\left(\sqrt{\frac{2}{\pi}}\frac{|a|\sqrt{N}}{\sigma}\right)\approx 1-Q\left(0.8\frac{|a|\sqrt{N}}{\sigma}\right), \tag{7.24}$$

which is slightly lower than that of the magnitude detection case. That is, more samples should be accumulated for the same probability of getting the correct update direction with the sign detection. Updating with error magnitudes is more reliable in predicting the update direction, and has other benefits such as faster convergence and smaller steady-state error. However, the advantage is not significant enough to justify the complexity over the simple update algorithm using signs only.

7.3.6 Complex Image Rejection

Image rejection is more accurate when performed with only complex signal and image without other blockers. Therefore, it is better to detect signs after *I*/*Q* channels are completely filtered. The error detector uses the *I*/*Q* outputs from the

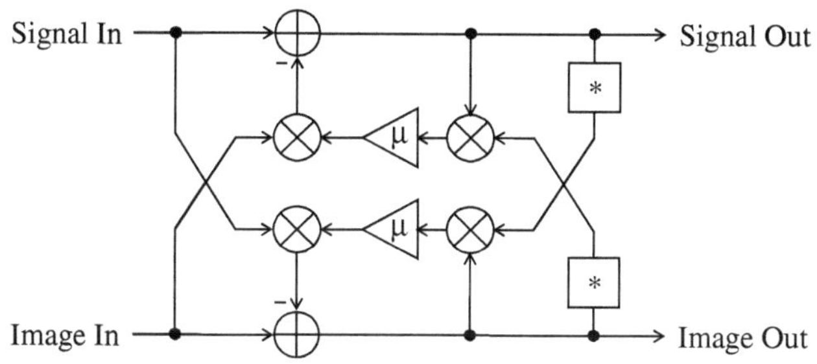

Fig. 7.20 Single-tap LMS complex image rejection concept

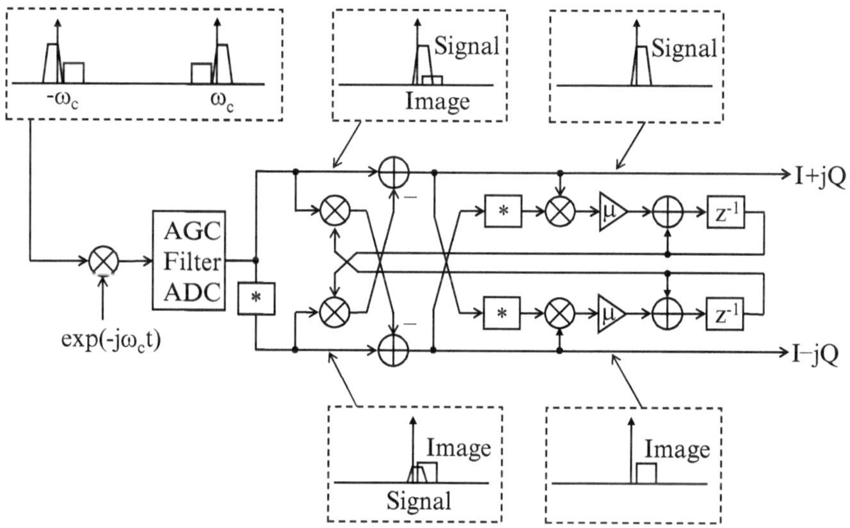

Fig. 7.21 Complex down-conversion receiver

image rejecter to estimate the gain and phase errors, and these are fed back to the image rejecter as shown in Fig. 7.20. Note that both the signal and the image are complex signal and its conjugate, $I+jQ$ and $I-jQ$, respectively, and * denotes the complex conjugate.

All baseband signals of direct-conversion receivers at zero- or low-IF are complex signals. Handling complex signals facilitates the image rejection. Complex image leaks into the complex signal, and complex signal leaks into the complex image. The leakage is symmetric with conjugate coefficients. The idea is to correlate the image out in the signal, and to subtract it from the signal. Similarly, the signal is correlated out to be subtracted in the image. Rather than calculating the image leakage for subtraction, it can be tracked continuously using an up/down counter-type register with the polarity of the image. It is basically the same LMS-based feedback system continuously detecting the image and minimizing it. The whole chain of the complex down-conversion receiver is sketched with the signal and image at different stages of the receiver in Fig. 7.21.

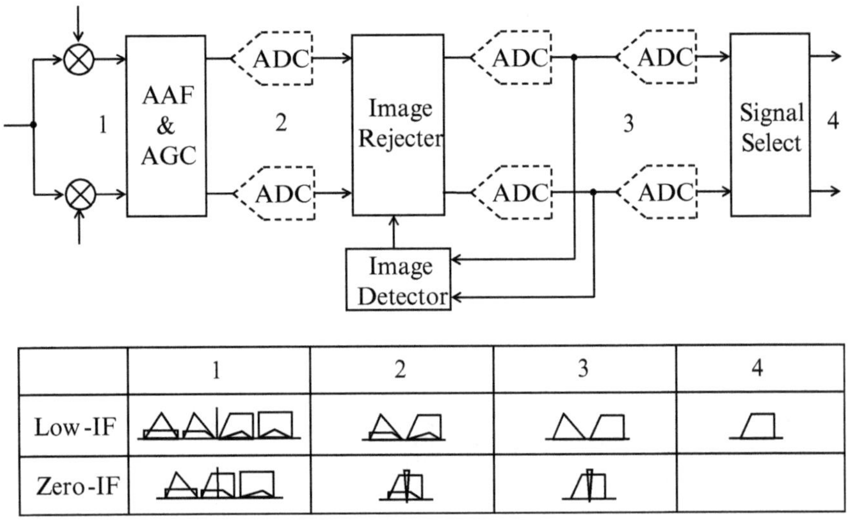

Fig. 7.22 Image rejection for low- and zero-IF receivers

Note the symmetry in the signal and image channels. The performance of the image rejecter depends on the location in the receiver where these signs are detected. *I*/*Q* ADCs can be located in different places for three possible variations of the implementation. The ADCs can be placed before the image rejecter for the completely digital implementation. The other extreme is the completely analog implementation in which both image rejecter and error detector are implemented in the analog domain. In the hybrid form, the image rejecter is made of an analog block, but the error detector can be implemented digitally. The ADCs, in this case, are placed after the image rejecter but before the error detector. Three ADC locations, and the complex *I*/*Q* spectra at three different points of the receiver are shown in Fig. 7.22.

Image rejection works well both in the zero- or low-IF receivers. However, the LMS algorithm accumulates error at DC. As a result, the accuracy of the error detector is affected by any signal and offset present at DC. For the reason, it requires that the DC component be notched out as marked. DC notching is not an issue in low-IF receivers. However, in the zero-IF case of receiving RF with fast-changing power envelope, DC calibration is a better solution than DC notching to avoid the well-known DC wandering problem.

7.3.7 *Three Variations of Image Rejecter*

The image rejection concept is much simpler and clearer if handled in the complex domain. However, real implementations start from two real *I*/*Q* signals. All digital implementation of the complex image rejection system is shown in Fig. 7.23.

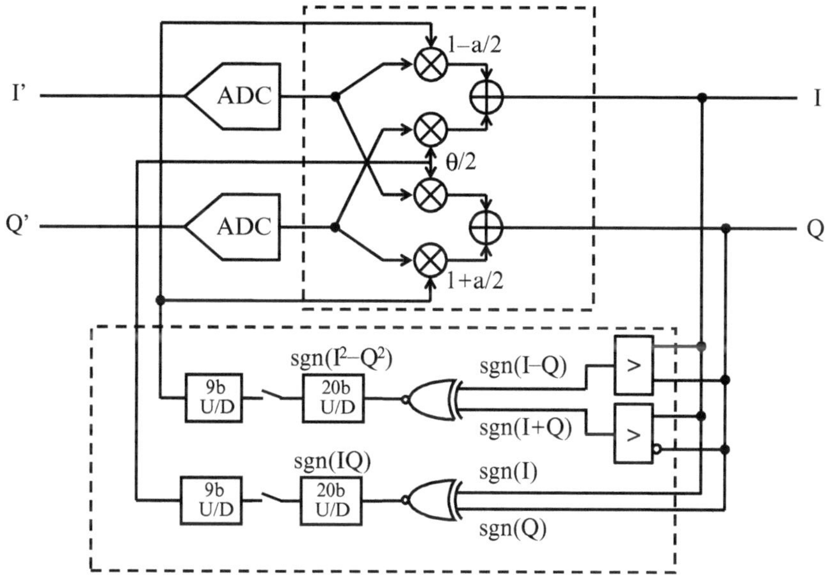

Fig. 7.23 All-digital image rejecter and detector

This symmetric image rejecter uses four multipliers and two adders, and the gains of the multipliers are constantly updated using the error detector. Note that an asymmetric image rejecter that corrects the I or Q path only while keeping the other path constant is also possible using two multipliers and one adder. Two digital comparators detect the signs of $I+Q$ and $I-Q$, and the signs of I and Q are determined by examining only the MSB sign bits. The products of these detected signs are fed into two 20b up/down (U/D) counters working as low-pass filters. The update directions of a and θ are decided after accumulating signs for 2^{20} cycles. This decision is used to increase or decrease the values stored in the 9b U/D counters. The advantage of the digital approach is that the sign detection is very accurate due to its high oversampling ratio, and the mismatch between the I/Q ADCs is corrected. The all-digital version has no other errors but the quantization error of the ADCs.

The analog and digital hybrid implementation is an alternative solution that is a compromise between the analog and the digital implementations. An analog sampled-data example of the image rejecter is a complex S/H with trim capacitors as shown in Fig. 7.24. This also serves as S/Hs for the following I/Q ADCs. The complex S/H should be linear, and its gain steps should be fine enough to achieve the desired IRR. Since the error detector is the same as in the digital case, this hybrid form has the benefits of the digital implementation. The mismatch of the I/Q ADCs is also corrected. The performance of the hybrid system is limited by the analog trimming step size. Therefore, only the system requirement sets the resolution of the ADCs.

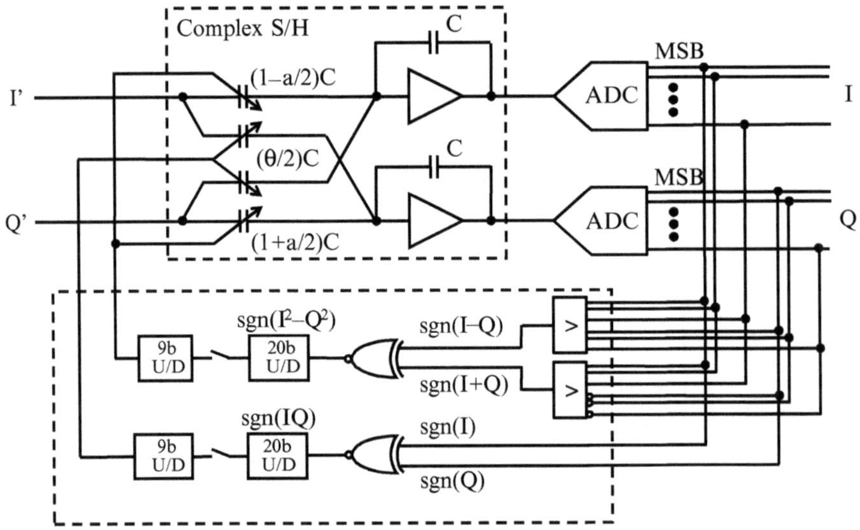

Fig. 7.24 Hybrid analog image rejecter and digital detector

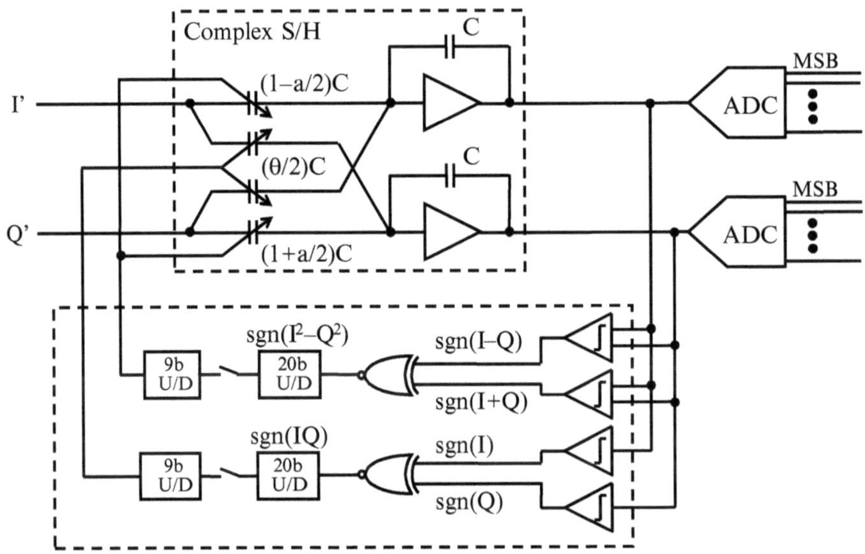

Fig. 7.25 All-analog image rejecter and detector

Figure 7.25 shows an all analog version of the image rejecter and detector. In the analog implementation using analog comparators, the image rejecter is also a complex *S*/*H* with trim capacitors, which is the same as in the hybrid system. The inputs of the error detector are the analog outputs of the *S*/*H*. These analog signals are latched using four comparators to get the signs of I, Q, $I+Q$, and $I-Q$. The rest of the error detector is the same as in the other two cases. Although high-resolution ADCs are not required as in the hybrid case, the disadvantage is that the IRR is limited by analog imperfections such as the offsets of analog comparators and the mismatch of the *I*/*Q* ADCs.

A complex switched-capacitor *S*/*H* circuit consists of two operational amplifiers, eight unit capacitors, and eight trim capacitors. The switches needed for operation are not shown. This block samples the *I*/*Q* inputs differentially on the bottom plates of two input capacitors that are digitally trimmed. Negative capacitance values can be obtained by reversing the input polarity in the differential implementation. The opamp gain should be set high enough to obtain better than 12b linearity. The *S*/*H* input capacitors are sized to be 1 pF so that the SNR limited by *kT*/*C* noise can be higher than 72 dB. The 9b trim capacitor can be made of a capacitor T-network with an array of 100 fF unit capacitors. Each trim capacitor covers a ±6.25 % range of the magnitude error with 0.024 % step and a ±3.5° range of the phase error with 0.014° step. If they are initialized to the middle values of the ranges, a total of 2^{20} times, 2^{8} cycles are needed to complete the worst-case initial adaptation assuming the step size is set to the minimum. However, the initial adaptation time can be shortened if the standard gear-shifting algorithm is used.

Analog comparator consists of two preamplifiers and a latch. The preamplifiers are implemented using a differential pair with diode-connected loads, and gain-enhanced using positive feedback. The latch is simply a dynamic positive-feedback latch. The offset and feed-through compensation is necessary since the input-referred offset of the comparator significantly affects the image rejection performance. The plot shown in Fig. 7.26 shows a simulation that quantifies the degradation of the IRR with input-referred offset of the comparator.

An offset as large as 1.5 mV for $1V_{pp}$ signal is acceptable for an IRR higher than 70 dB. Two preamplifiers and capacitors are switched for offset cancellation based

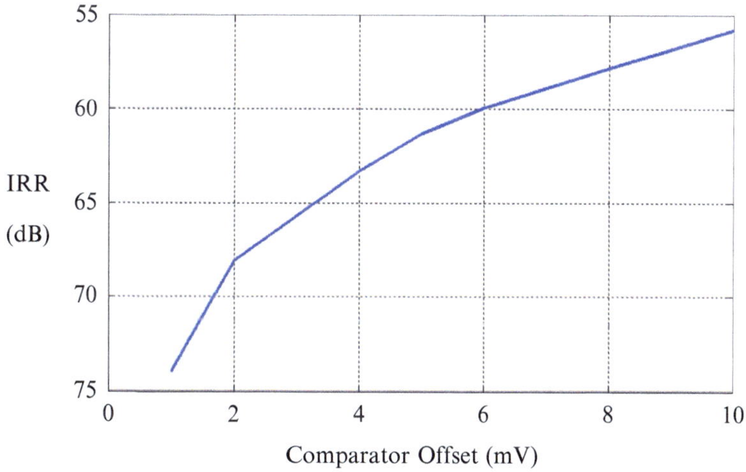

Fig. 7.26 IRR affected by the offset of analog polarity detector

on the correlated double sampling principle. The offset and feed-through error of the first preamplifier is amplified, and sampled on the input capacitor of the second preamplifier stage, thereby reducing the first-stage offset and feed-through error.

7.4 Experimental Results

The *I*/*Q* signals of $1V_{pp}$ are generated at 1 MHz IF at an 800 kHz symbol rate. The complex image blocker used for testing has the same power level as the complex signal for all the IRRs to be relative to the signal level. The signals sampled at 40 MS/s are digitized at 5 MS/s using two 15b ADCs. The measured complex spectra for all three cases are shown in Fig. 7.27.

The IRR, before image rejection, is about 26 dB with a gain mismatch of 5 % and phase mismatch of 3°. The image is rejected by about 65 dB in the digital implementation using 12b ADCs. For the hybrid case, the image is also rejected by about 65 dB. The analog case, however, exhibits a lower IRR of 62 dB mainly due to the offsets of analog comparators. With 15b ADCs in the digital implementation, the image is suppressed well below the noise level of the test set-up.

The image of a strong blocker is suppressed all the way down to the quantization noise level. As long as the blocker stays within the ADC range, its image is rejected. That is, the achievable IRR varies depending on the image power. The signal shown

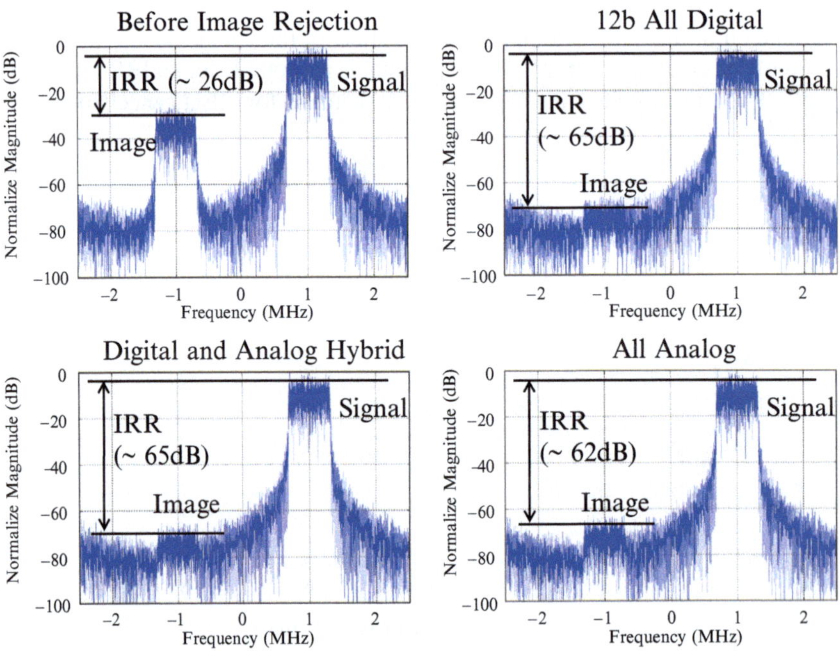

Fig. 7.27 Image spectra before and after image rejection

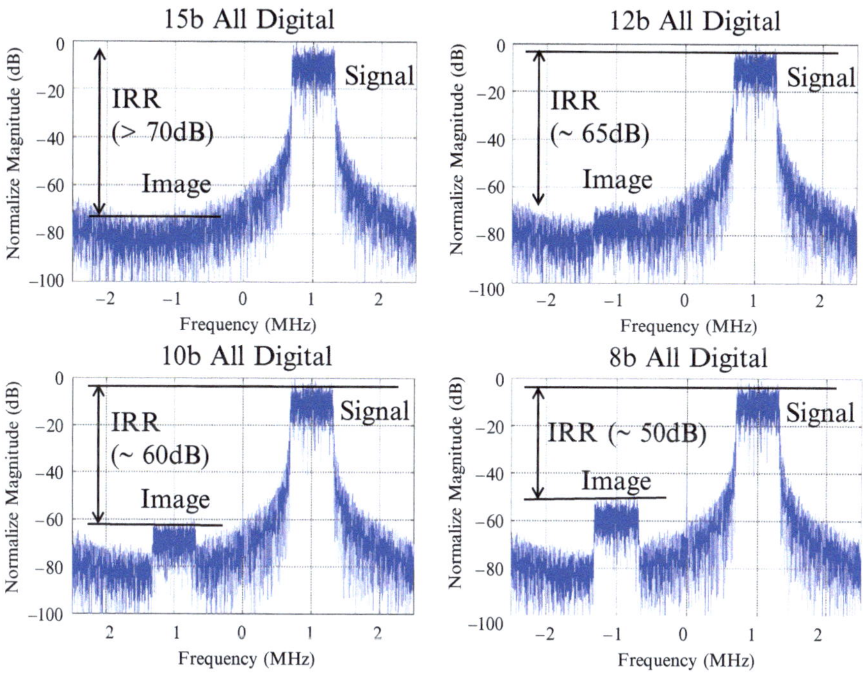

Fig. 7.28 IIR vs. the number of bits

is a zero-IF complex signal, which happens to have the signal on the positive side of the band only and its own image falls on the negative side. Or it can be considered as a low-IF complex signal with an image of the same magnitude blocker. The image rejection principle can be equally applied to both zero- and low-IF cases. In digital wireless receivers, the trend is to use high SFDR ADCs for digital channel filtering and global AGC, but most RF systems cannot afford to have wide dynamic range ADCs. With lower number of bits, IRRs of only 60 and 50 dB are measured with 10b and 8b ADCs as shown in Fig. 7.28.

Figure 7.29 shows the measured 256-QAM constellation before and after image rejection. The constellation is significantly dislocated from its ideal location with an IRR of 26 dB before image rejection. It can be seen that the constellation goes back to the ideal location after image rejection by 65 dB. The measured eye diagram of the 256-QAM signal shows that the eyes are almost closed before image rejection but they open up completely after image rejection as shown in Fig. 7.30.

The error probabilities vs. SNR for 64- and 256-QAM signals are plotted as a function of IRR in Fig. 7.31. As shown, the image is the dominant source of performance degradation for high SNR constellations. Poor IRR causes the dramatic increase in the BER of the received data. This shows that the 256-QAM case is more sensitive to image than the 64-QAM case. In addition, image is a more serious problem when the SNR is high. With an IRR > 60 dB, the error probability converges to the ideal value but the error probability increases very rapidly with an IRR < 60 dB.

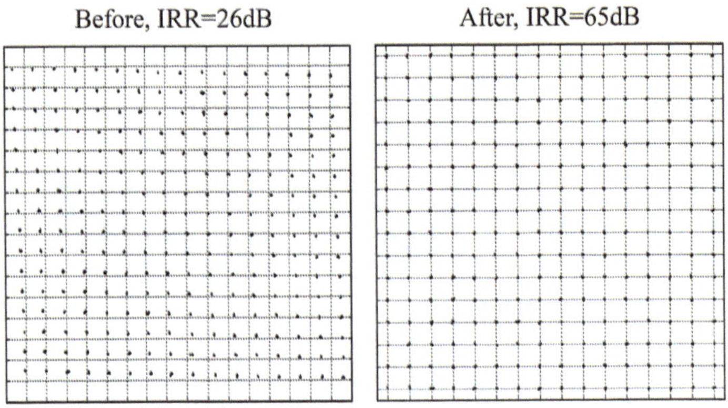

Fig. 7.29 256-QAM constellation before and after image rejection

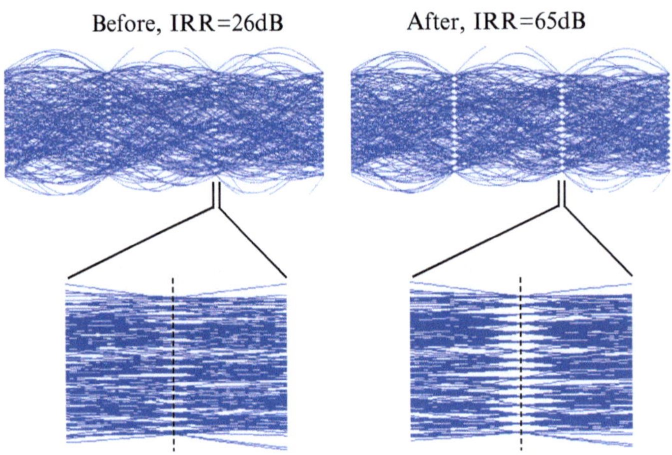

Fig. 7.30 Eye diagram of 256-QAM before and after image rejection

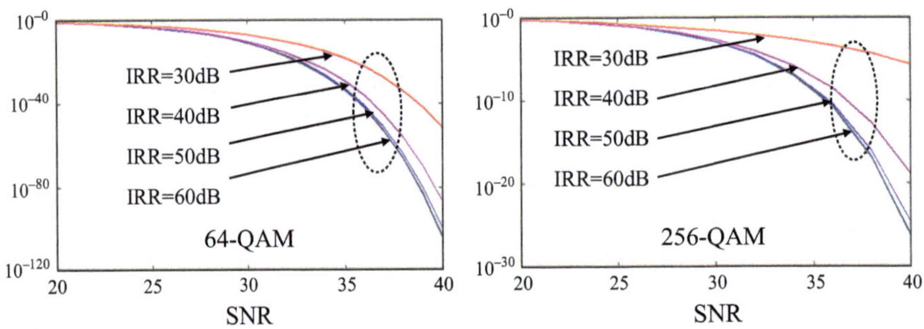

Fig. 7.31 Error probability of 64- and 256-QAM vs. IRR

One good example of RF receivers that require very high-level harmonic and image rejections is a broadband TV tuner. TV channels cover a broad range of spectrum over four octaves from VHF to UHF. The harmonic mixing and image are most critical design constraints in TV tuners. Harmonic channels can be somewhat filtered with a roughly tuned resonator, and suppressed by poly-phase harmonic mixer. The low-IF image is rejected in the baseband after the complex signal is obtained.

Figure 7.32 shows a direct-conversion low-IF TV tuner example with both harmonic and image rejections [17]. The image-rejected complex signal is converted back to the real IF using a Hilbert filter. With one resonator tuned to RF, the third-harmonic and the image channels are suppressed by 72 and 61 dB, respectively as shown in Fig. 7.33.

Figure 7.34 shows that the luminance signal of the old adjacent analog NTSC channel folded into the baseband in this low-IF receiver with 40 dB suppressed. It is rejected by 60 dB with the image rejection algorithm turned on. Digital cable TV

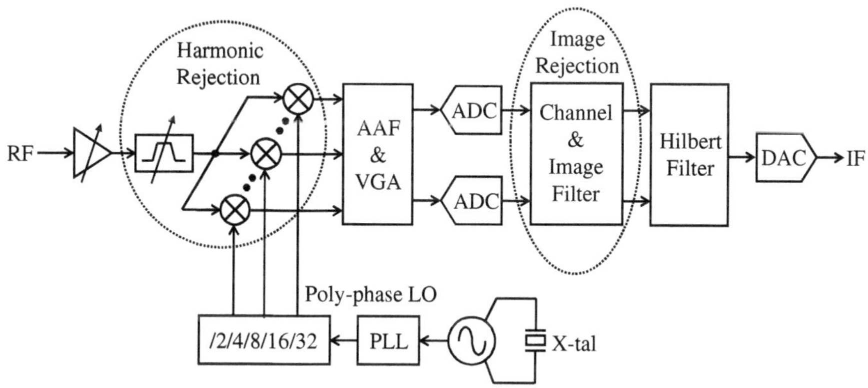

Fig. 7.32 Low-IF TV tuner example

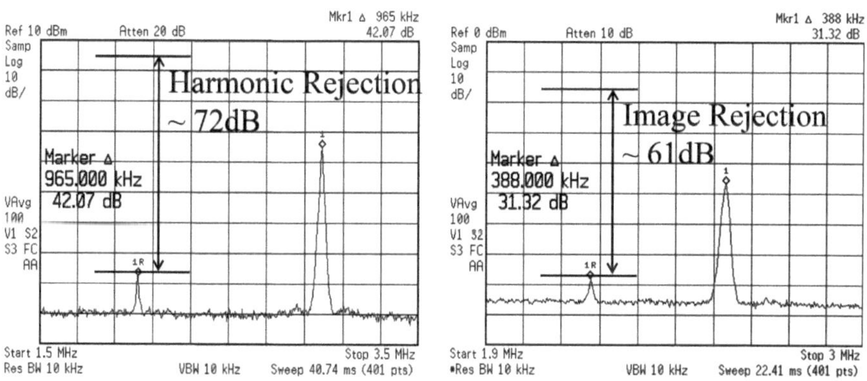

Fig. 7.33 Harmonic and Image Rejection in the down-conversion

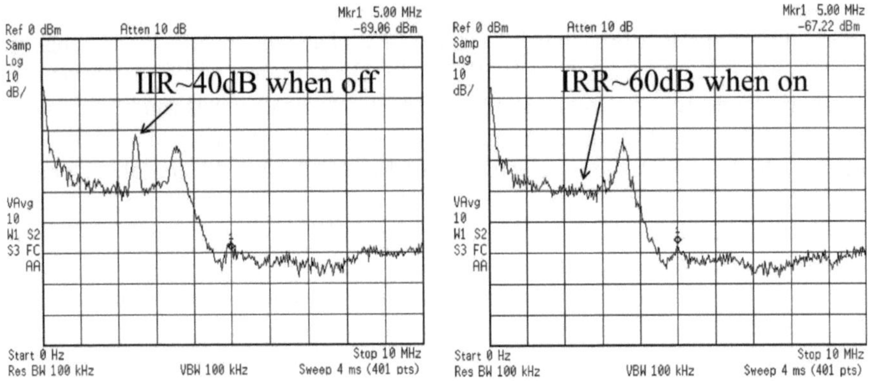

Fig. 7.34 Analog TV tuner image

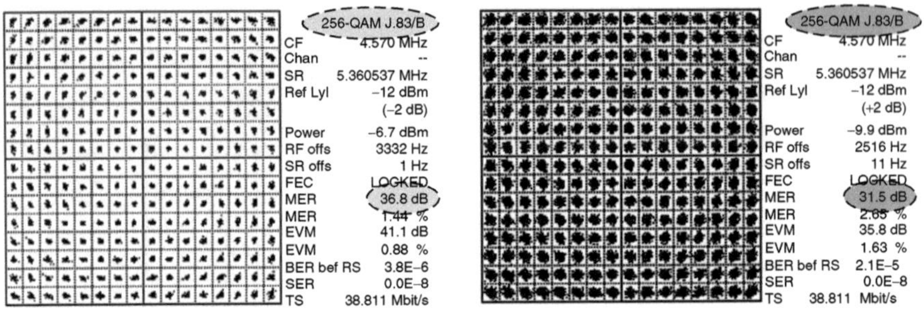

Fig. 7.35 Actual digital cable TV channel with 256-QAM

uses 256-QAM. On the left side of Fig. 7.35, the source signal with a modulation error rate (MER) of 36.8 dB is shown, and on the right side, the received and demodulated constellation is shown with an MER of 31.5 dB.

Image frequencies corrupting the down-converted desired signal result from the gain and phase errors of *I*/*Q* signals. The image problem has been one of the major obstacles to overcome when implementing direct-conversion low- or zero-IF receivers. A post-processing image rejection algorithm can detect and correct *I*/*Q* imbalance continuously. The algorithm is based on an adaptive zero-forcing feedback concept using sign bits only, and does not require complicated digital processing. It is basically the same first-order 1b $\Delta\Sigma$ modulator feedback with high oversampling ratio. It is to trim the constant DC image parameter based on the DC servo feedback principle.

The total accumulated image component at the system level is rejected in the complex baseband regardless of the source of the imbalance. The image rejecter and error detector are implemented either in the analog or the digital domain. The hybrid implementation with a complex *S*/*H* and digital sign detectors alleviates the need for a high-resolution ADC in the digital image rejecter. The image rejection algorithm facilitates simple zero- or low-IF RF receiver implementations since no accurate analog circuits are necessary for complex I/Q processing. Due to its simplicity, the digital implementation of the image rejecter is preferred to the analog version. The analog/digital hybrid implementation can be made just as good without using high-resolution ADCs.

References

1. F. Behbahani, Y. Kishigami, J. Leete, A. Abidi, CMOS mixers and polyphase filters for large image rejection. IEEE J. Solid State Circuits **36**, 873–887 (2001)
2. S. Jantzi, K. Martin, M. Snelgrove, A. Sedra, Complex bandpass $\Delta\Sigma$ converter for digital radio, in *IEEE International Symposium on Circuits Systems*, vol. 5, May 1994, p. 453
3. J. Crols, M. Steyaert, An analog integrated polyphase filter for a high performance low-IF receiver, *VLSI Dig. Tech. Papers*, Jun. 1995, pp. 848–854
4. A. Rofougaran, J. Rael, M. Rofougaran, A. Abidi, A 900MHz CMOS LC-oscillator with quadrature outputs, *ISSCC Dig. Tech. Papers*, Feb. 1996, pp. 392–393
5. L. Der, B. Razavi, A 2GHz CMOS image-reject receiver with LMS calibration. IEEE J. Solid State Circuits **38**, 167–175 (2003)
6. M. Elmala, S. Embabi, Calibration of phase and gain mismatches in weaver image-reject receiver. IEEE J. Solid State Circuits **39**, 283–289 (2004)
7. J. Glas, Digital I/Q imbalance compensation in a low-IF receiver, in *Global Telecommunications Conference*, Nov. 1998, pp. 1461–1466
8. K. Pun, J. Franca, C. Azeredo-Leme, A digital method for the correction of I/Q phase errors in complex sub-sampling mixers, in *Proceedings of 2000 Southwest Symposium on Mixed-Signal Design*, Feb. 2000, pp. 171–174
9. M. Dawkins, A. Burdett, N. Cowley, A single-chip tuner for DVB-T. IEEE J. Solid State Circuits **38**, 1307–1317 (2003)
10. L. Yu, W.M. Snelgrove, A novel adaptive mismatch cancellation system for quadrature IF radio receivers. IEEE Trans. Circuits Syst. II **46**(6), 789–801 (1999)
11. C. Heng, M. Gupta, S. Lee, D. Kang, B. Song, A CMOS TV tuner/demodulator IC with digital image rejection. IEEE J. Solid State Circuits **40**, 2525–2535 (2005)
12. I. Elahi, K. Muhammad, P. Balsara, I/Q mismatch compensation using adaptive decorrelation in a low-IF receiver in 90-nm CMOS process. IEEE J. Solid State Circuits **41**, 395–404 (2006)
13. S. van Gervan, D. van Compernolle, Signal separation by symmetric adaptive decorrelation: stability, convergence, and uniqueness. IEEE Trans. Signal Process. **43**(7), 1602–1612 (1995)
14. B. Widrow et al., The complex LMS algorithm, Proc. IEEE **63**(4), 719–720 (1975)
15. S. Lerstaveesin, B. Song, A complex image rejection circuit with sign detection only, *ISSCC Dig. Tech. Papers*, Feb. 2006, pp. 454–455
16. S. Lerstaveesin, B. Song, A complex image rejection circuit with sign detection only. IEEE J. Solid State Circuits **41**, 2693–2702 (2006)
17. S. Lerstaveesin, M. Gupta, D. Kang, B.S. Song, A 48-860MHz CMOS low-IF direct conversion DTV tuner. IEEE J. Solid State Circuits **43**, 2013–2024 (2008)

MIX
Papier aus verantwortungsvollen Quellen
Paper from responsible sources
FSC® C105338

If you have any concerns about our products,
you can contact us on
ProductSafety@springernature.com

In case Publisher is established outside the EU,
the EU authorized representative is:
Springer Nature Customer Service Center GmbH
Europaplatz 3, 69115 Heidelberg, Germany

Printed by Libri Plureos GmbH
in Hamburg, Germany